THERMAL EXPANSION 8

A Continuation Order Plan is available for this series. A continuation order will bring delivery of each new volume immediately upon publication. Volumes are billed only upon actual shipment. For further information please contact the publisher.

THERMAL EXPANSION 8

Edited by

Thomas A. Hahn
Naval Research Laboratory
Washington, D.C.

SPRINGER SCIENCE+BUSINESS MEDIA, LLC

Library of Congress Cataloging in Publication Data

International Thermal Expansion Symposium (8th: 1981: Gaithersburg, Maryland)
Thermal expansion 8.

"Proceedings of the Eighth International Thermal Expansion Symposium, held June 15–17, 1981, at the National Bureau of Standards, Gaithersburg, Maryland."—T.p. verso.

Includes bibliographical references and index.

1. Expansion (Heat)—Congresses. I. Hahn, Thomas, A. II. Title.

QC281.5.E9158 1981 536'.41 84-17820
ISBN 0-306-41825-8

ISBN 978-1-4899-5006-2 ISBN 978-1-4899-5004-8 (eBook)
DOI 10.1007/978-1-4899-5004-8

Proceedings of the Eighth International Thermal Expansion Symposium, held June 15–17, 1981, at the National Bureau of Standards, Gaithersburg, Maryland

Originally published by Plenum Press, New York in 1984
Softcover reprint of the hardcover 1st edition 1984

Dedicated to the memory of
Dr. Y. S. Touloukian

FOREWORD

The International Thermal Expansion Symposia were started in 1968 at the initiative of R.K. Kirby and P.S. Gaal. These Symposia cover the developments and advances in theoretical and experimental studies of the thermal expansion of solids and its relation to other related properties, and provide a broadly based forum for researchers actively working in this field to convene on a regular basis to exchange ideas and experiences and report their findings and results.

The Symposia have been self-perpetuating and are an example of how a technical community with a common purpose can transcend the artificial barriers between disciplines and gather together in increasing numbers without the need of national publicity and continuing funding support, when they see something worthwhile going on.

Of the first five Symposia, only three published formal Proceedings. However, beginning with the Sixth Symposium in 1977, when the Center for Information and Numerical Data Analysis and Synthesis (CINDAS) of Purdue University became the permanent sponsor of the Symposia, a policy of publishing formal Proceedings on a continuing and uniform basis was established. Thus, including the present volume, the following formal Proceedings have been published:

Symposium and Year	Title of Volume	Publisher and Year
2nd (1970)	SYMPOSIUM ON THERMAL EXPANSION OF SOLIDS - Journal of Applied Physics, 41(13), pp. 5043-5154	American Institute of Physics (1970)
3rd (1971)	THERMAL EXPANSION - 1971 AIP Conference Proceedings No. 3	American Institute of Physics (1972)
4th	THERMAL EXPANSION - 1973 AIP Conference Proceedings No. 17	American Institute of Physics (1974)
6th (1977)	THERMAL EXPANSION 6	Plenum Press (1978)
7th (1979)	THERMAL EXPANSION 7	Plenum Press (1982)
8th (1981)	THERMAL EXPANSION 8	Plenum Press (1984)

Mr. Thomas A. Hahn, General Chairman of the Eighth Symposium, is to be congratulated for his excellent leadership in conducting the Symposium and for his painstaking efforts, which made the present volume possible. CINDAS looks forward to working with future host institutions to ensure that future Symposia continue to produce high-quality volumes of Proceedings in this important, specialized field.

This Foreword should have been written by Dr. Y.S. Touloukian, the founder and founding Director of CINDAS for 25 years. It was owing to Dr. Touloukian's great efforts that the Proceedings of recent (and future) Symposia have been (and will be) published formally on a continuing and uniform basis, thus making them a major permanent vehicle for the reporting of research results on thermal expansion. His passing away at the time of this Symposium is a great loss to the field of thermophysical properties, to which he had devoted his entire life.

C.Y. Ho

Director
Center for Information and Numerical Data Analysis and Synthesis
Purdue University

West Lafayette, Indiana
January 1984

PREFACE

The 8th International Thermal Expansion Symposium (ITES) was held at the National Bureau of Standards, Gaithersburg, Maryland on June 15-17, 1981. The joint sponsors of the symposium were NBS and the Center for Information and Numerical Data Analysis and Synthesis (CINDAS) of Purdue University. The general chairman was Thomas A. Hahn, then with NBS and currently with the Naval Research Laboratory, Washington, D.C.

The 8th ITES was held concurrently with the 8th Symposium on Thermophysical Properties (J.V. Sengers, Chairman) and the 17th International Thermal Conductivity Conference (J.G. Hust, Chairman). Overall coordination of this joint conference was provided by the International Thermophysical Congress (A. Cezairliyan, Chairman). The plenary session of the joint conference was welcomed by J.D. Hoffman (NBS), S. Gratch (American Society of Mechanical Engineers), and J. Kestin (Brown University), who gave the keynote address on Thermophysical Properties in Science and Technology.

Special appreciation is due to the many people who made the 8th ITES possible, including the organizing committee of the joint conference; the arrangements committee of NBS, including R.F. Martin, J.A. Lorden, and K.C. Stang; the ITES Governing Board; and the session chairmen. The work of the session chairmen in reviewing the manuscripts for this volume is greatly appreciated. The support of the Naval Research Laboratory and the efforts of P. Ridgeway and J. Dalton of NRL have made the preparation of this volume possible.

PREFACE

[illegible]

CONTENTS

SESSION 1: THEORY

Chairman: T.A. Hahn

SESSION 2: OXIDES

Chairman: W.A. Plummer

SESSION 5: PROCESSING

Chairman: P. Gaal

SESSION 6: REFERENCE TECHNIQUES

Chairman: G.K. White

VARIATIONS OF GRÜNEISEN PARAMETER γ FOR NON-METALLIC CRYSTALS

Guy K. White

CSIRO Division of Applied Physics
Sydney, Australia 2070

INTRODUCTION

This paper reviews briefly, from an experimental viewpoint, the present state of knowledge of the lattice vibrational contribution to the thermal expansion for various classes of crystalline solids. A fuller review[1] and book[2] include a discussion of experimental methods, electronic and magnetic contributions, the effect of impurities, and theoretical models.

Firstly, a reminder of the thermodynamic relations that link the volume expansion coefficient β, bulk modulus B, heat capacity C and pressure P, volume V, temperature T, etc.: $\beta = (\partial P/\partial T)_V/B_T$ emphasizes that β depends both on the change in internal pressure generated by change in T at constant volume <u>and</u> on the elastic stiffness, B_T, of the solid.

The dimensionless Grüneisen function $\gamma = \beta V B_T/C_V \equiv \beta V B_S/C_P$ relates the <u>two</u> strongly temperature-dependent quantities β and C but is itself relatively weakly dependent on temperature.

Equivalent definitions are

$$\gamma = -\left(\frac{\partial \ln T}{\partial \ln V}\right)_S = V\left(\frac{\partial P}{\partial T}\right)_V/C_V = \left(\frac{\partial S}{\partial \ln V}\right)_T/C_V. \quad \text{(S is lattice entropy.)}$$

γ raises its ugly head in many physical situations but it is not always the same average of all the microscopic γ_i each of which expresses the volume dependence of an individual mode of vibration of frequency ω_i in the equation

$$\gamma_i = -\,d\ln\omega_i/d\ln V.$$

At very low temperatures where only long waves are excited, the Debye continuum model becomes valid and lattice frequencies can be calculated from a knowledge of the elastic moduli, c_{ij}. Thermal measurements will give a value γ_o^{th} which should agree with $\gamma_o^{el} = - d\ln\theta_o/d\ln V$ calculated from suitable averaging of the pressure dependences of the elastic constants. Here θ_o is the low temperature limiting value of the Debye temperature θ_D. At temperatures high compared with θ_D all modes are excited and $\gamma_\infty = \langle \gamma_i \rangle$, including optic modes.

The review[1] indicates some of the so-called Grüneisen parameters which are frequently confused. These include the approximations of Slater, Dugdale and MacDonald and the quite distinct Anderson-Grüneisen function.

The average quantity γ^2 or $\langle \gamma_i^2 \rangle$ is important in determining the thermal conductivity of non-metals where the salient modes are less certain. In a recent review Slack[3] has shown that use of γ_∞ in the modified form of the Leibfried-Schlömann equation gives values for heat conductivity γ which are usually within a factor of 2 of the experimental value for a large number of solids of rock-salt and zincblende structure.

Below I shall review how γ^{th} varies with T for various crystals and mention briefly the dominant vibrational modes where they are known.

CUBIC CRYSTALS

Rare gas solids. Swenson and his colleagues[4,5] were the first to measure the linear coefficient $\alpha(T)$ to liquid helium temperatures. Figure 1 shows the relatively small variation in $\gamma(T)$

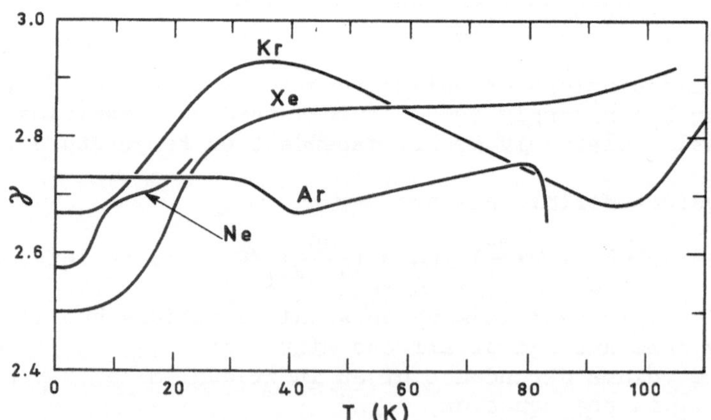

Fig. 1. Grüneisen parameter for rare gas solids.[4,5]

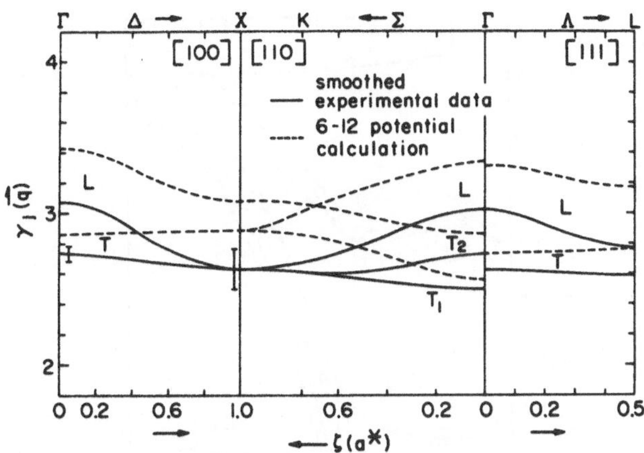

Fig. 2. Neon. Variation of γ_i with wave number in three symmetry directions

from 1.5 K to the melting point which Barron[6] had predicted would be the case for a nearest-neighbour close-packed model.

For neon, $\gamma_o^{th} \approx 2.58$, which agrees well with $\gamma_o^{el} = 2.60 \pm 0.02$ calculated from neutron scattering measurements on solid neon at 0 and 6 kbar pressure. Other deductions from the neutron studies are that γ_i values for LA and two TA branches are very similar and do not change significantly with wave number showing why the thermally observed $\gamma(T)$ does not vary much with T (Fig. 2).

The NaCl structure. Among the alkali halides, the picture is very different: γ_i is highly anisotropic particularly for the K- and Rb-halides and indeed negative for the (100)-or c_{44}-type shear modes. These are also of the lowest frequency and so are heavily weighted at low temperatures. Measurements[1] (Fig. 3) show that γ_∞ lies between 1.4 and 1.7 for all the halides. For LiF $\gamma(T)$ remains almost constant and $\gamma_o^{el} = 1.65$. γ falls around $\theta/10$ to $\gamma_o \sim 1.0$ for Na halides, to ~ 0.3 for K-halides, and to 0 or -0.1 for Rb-halides. Within limits of error γ_o^{th} agrees with γ_o^{el} (solid bars in Fig. 3) and the values reflect the decrease in c_{44} shear stiffness and the fact that these modes become softer, i.e. $d\omega/dP$ becomes negative as the alkali ion increases in diameter.

For the alkaline earth oxides, there are less thermal data but elastic measurements show that γ_o is 1.6 for MgO and falls to 0.8 for BaO.[1] This is associated with the same shear modes as in alkali halides, for which γ_i is positive in MgO and CaO but negative in SrO and BaO.

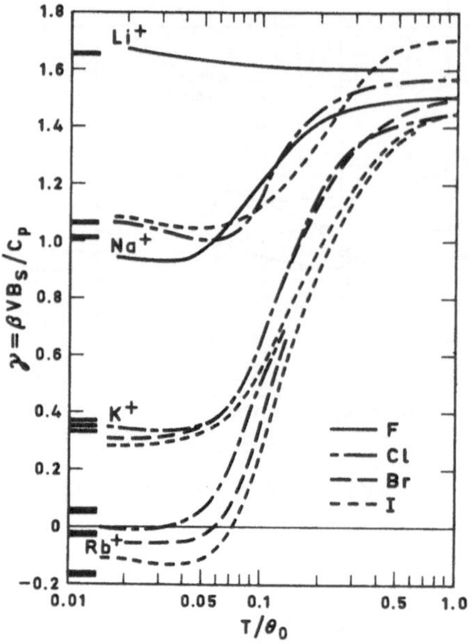

Fig. 3. γ(T) for alkali halides of rock-salt structure.[1]

<u>The CsCl structure</u>. The caesium halides show much less anisotropy in the elastic moduli, no soft modes, and are almost perfect 'Grüneisen' solids with $\gamma \approx 2.0$ from 2 to 300 K.[1]

<u>The fluorite structure</u>. For the alkaline earth fluorides, as we progress from Ca to Sr to Ba, the ion size increases and we find a pattern similar to that in the rock-salt family except that it is the shear stiffness $c' = \frac{1}{2}(c_{11}-c_{12})$ in (110) directions that decreases and becomes soft: $\gamma_0 \sim 1.1$ for CaF_2, 0.7 for SrF_2, and 0.3 for BaF_2 (Fig. 4). Included are ThO_2[8] and PbF_2.[9] For the latter $\gamma_0 \approx 0.8$ but $\gamma(T)$ has a marked maximum at 25 K associated with an unidentified optic mode.

<u>The zincblende or diamond structure</u>. In this tetrahedrally-bonded family there are also low-lying TA modes controlled by the shear stiffness c' which can soften under pressure. Novikova and her collaborators[10,1] first showed that most of these solids have negative coefficients at low temperatures. Since then many have been measured to liquid helium temperatures. $\alpha(T)$ and $\gamma(T)$ become increasingly negative as we progress from group IV (Ge, Si) to III-V, to II-VI and finally to a I-VII compound, CuCl, for which $\gamma_0 \sim -2$ (Fig. 5). The progressive decrease in γ_0 and γ_{min} is associated with increasing ionicity of the bonding (and less angular rigidity), decrease in shear stiffness c' and increased softening of c' under pressure. The deep minimum at $\theta/15$ cannot

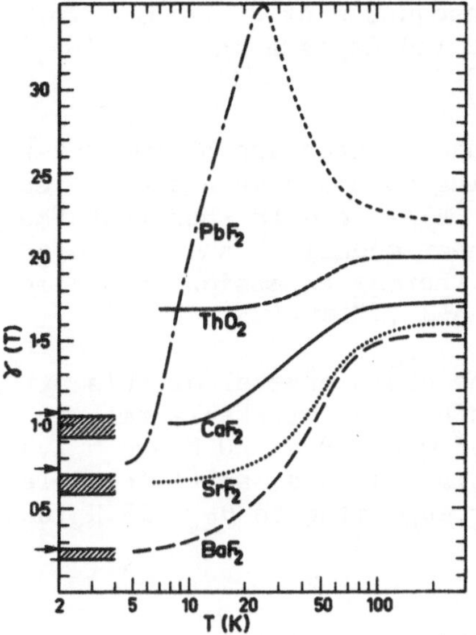

Fig. 4. γ(T) for alkaline earth fluorides,[1,9] PbF_2[9] and ThO_2.[8]

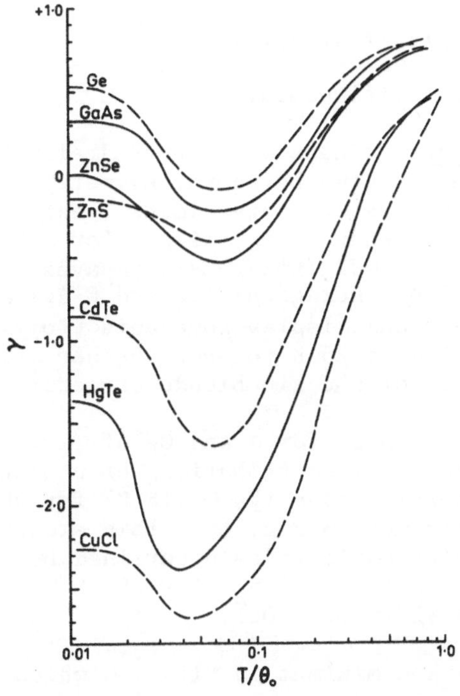

Fig. 5. γ(T) for zincblende structure.[1]

arise solely as a consequence of dispersion for this mode as γ_i (q = 0) > γ_{min} but indicates a decrease in γ_i towards the zone boundary.

Cuprite. Cu_2O has a structure of two interpenetrating zincblende lattices sharing common oxygen atoms; each O is surrounded tetrahedrally by Cu. Elastic data show that the principal TA modes controlled by both shear moduli c' and c_{44} soften under pressure with γ_i values of *ca* -4. Thermal expansion measurements[11] show β is negative below 270 K and $\gamma_o^{th} \approx -4.5$.

Perovskites. One cubic crystal of this structure, $KTaO_3$, has been measured and exhibits a remarkable rise in γ at low temperatures: $\gamma \approx 45$ at 4 K compared with 19 at 20 K and ~ 1.3 at room temperature.[12] This is attributed to a 'soft' ferroelectric mode with an excitation energy corresponding to $\theta_E \sim 13$ K (near 3 K).

ANISOTROPIC CRYSTALS

For those crystals with axial symmetry, there are two principal coefficients of linear expansion $\alpha_{\parallel}$ (parallel) and $\alpha_{\perp}$ (perpendicular) and two principal Grüneisen functions given by

$$\gamma_{\parallel} = [c_{33}\alpha_{\parallel} + 2c_{13}\alpha_{\perp}]\ V/C$$

$$\gamma_{\perp} = [(c_{11} + c_{12})\alpha_{\perp} + c_{13}\alpha_{\parallel}]\ V/C$$

where c_{ij} are elastic stiffnesses.

Wurtzite structure. These are tetrahedrally bonded like the zincblende structure but have hexagonal symmetry. Although thermal data at the lowest temperatures are limited, extrapolation indicates that $\gamma_{\parallel}$ and $\gamma_{\perp}$ for ZnO behave very like γ for ZnS (cubic) in form with $\gamma_o^{el} \approx -1.1$[13]; also CdS (hexagonal) behaves very like CdTe (cubic) with $\gamma_o \approx -1.4$. Hexagonal ice and β-AgI also have negative expansion coefficients and display some anisotropy. Measurements are not yet available below *ca* 25 K to show whether γ(T) has the deep minimum characteristic of the zincblende structure (see review[1]).

Rutile structure. Data for α and C_P of TiO_2 at low temperatures may be analysed in terms of contributions of acoustic modes ($\theta_o \approx 770$ K) and Einstein modes ($\theta_E \approx 118$ K) for which $\gamma_E \approx 7.5$, producing a marked maximum in Fig. 6. This excitation energy does not correlate well with optic modes determined by spectroscopy, e.g. with B_{1u}, B_{1g} or the 'soft mode' A_{2u} for which $\gamma_i \approx 13$. Two sets of elastic data lead to γ_o^{el} = 0.4, 0.5.

MgF_2 shows a marked minimum in γ(T)[1,14] which could arise from an optic mode of $\theta_E \approx 155$ K. No pressure data on optic modes are available but elastic data lead variously to γ_o^{el} = 0.3, 0.6.

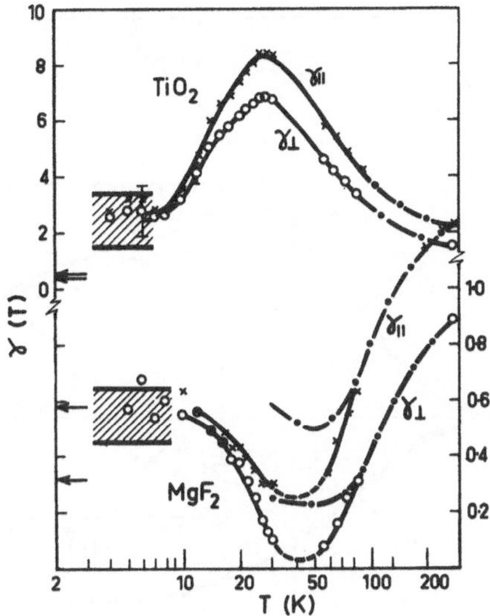

Fig. 6. Rutile structure: $\gamma(T)$ for TiO_2 and MgF_2.

α-quartz. This has trigonal symmetry and shows considerable anisotropy in α and γ values at low temperatures. As $T \to 0$, $\gamma_\perp \approx +0.8$ while $\gamma_\parallel \approx -0.6$, characteristic of chain structures (see below) and may be attributed to the tight -Si-O-Si-O helices with their axes in the c-direction.[1]

Se, Te and polyoxymethylene $(CH_2O)_n$. These are three chain-like crystals for which there are data to low temperatures.[1] In each there is strong covalent bonding along chains and weak van der Waals forces between them. Thus, they are highly compressible normal to the chain-axes but stiff along the chains. At all temperatures, α is highly anisotropic with $\alpha_\perp$ relatively large and positive while $\alpha_\parallel$ is small or negative. The two principal γ-values are similar at high temperature but in the low temperature limit where only low-frequency modes are important, $\gamma_\perp$ is positive and $\gamma_\parallel$ is negative (Fig. 7). Physically this means that lowest frequencies are increased by stretching along the axis and are weakened by stretching normal to the axes.

Graphite and boron nitride. In these layer-structure crystals the opposite situation occurs: they are soft in the c-direction so that $\alpha_\parallel$ is positive while $\alpha_\perp$ is negative below room temperature.[16,1] At low temperatures $\gamma_\parallel \approx 1$ while $\gamma_\perp$ is negative.

SUMMARY

The variation of γ with T for a crystal reflects both the spread of γ_i among various modes of vibration and the change in

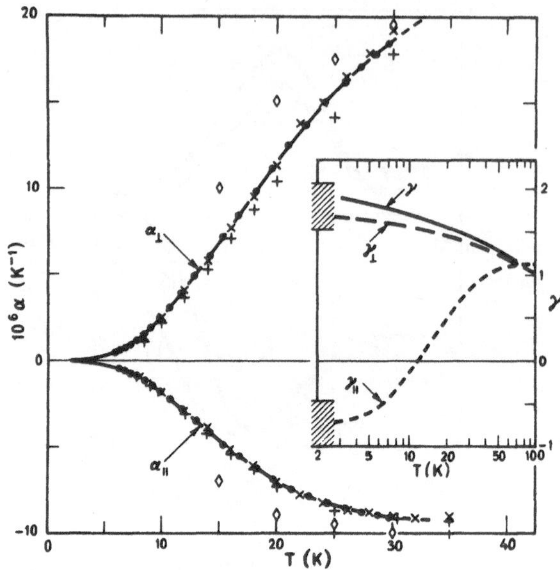

Fig. 7. $\alpha(T)$ and $\gamma(T)$ (inset) for Te.[15]

importance (or weighting) of these modes. Close-packed cubic crystals of the rare gases are like most cubic metals in that γ_i is positive for all modes and varies only between, say, +2 and +3. Although TA modes are more important at low temperatures, the weighted average γ_o is not very different from γ_∞. However for the structures of lower coordination number, such as rocksalt and zincblende, γ_i can vary greatly with direction, polarization and frequency and those modes with small or negative values may be heavily weighted at low temperatures resulting in γ_o being small or negative.

There may also be optic modes which are strongly strain-dependent and these will affect the average $\langle \gamma_i \rangle$ in the vicinity of their excitation temperature, θ_E. Examples are $KTaO_3$ at low temperatures, PbF_2 at intermediate temperatures, and ZnS at high temperatures.

ACKNOWLEDGMENT

I thank my colleague, Dr J.G. Collins, for his frequent and generous assistance.

REFERENCES

1. T.H.K. Barron, J.G. Collins, and G.K. White, Thermal expansion of solids at low temperatures, Adv. Phys. 29:609 (1980).
2. R.S. Krishnan, R. Srinivasan, and S. Devanarayanan, "Thermal Expansion of Crystals", Pergamon, Oxford (1979).
3. G.A. Slack, Thermal conductivity of non-metallic crystals, Solid State Phys. 34:1 (1979).
4. C.R. Tilford and C.A. Swenson, Thermal expansions of solid argon, krypton and xenon above 1 K, Phys. Rev. B 5:719 (1972).
5. J.C. Holste and C.A. Swenson, Experimental thermal expansion for solid neon, 2-14 K, J. Low Temp. Phys. 18:477 (1975).
6. T.H.K. Barron, On the thermal expansion of solids at low temperatures, Phil. Mag. 46:720 (1955).
7. J. Eckert, W.B. Daniels and J.D. Axe, Phonon dispersion and mode Grüneisen parameters in neon at high density, Phys. Rev. B 14:3649 (1976).
8. G.K. White and F.W. Sheard, Thermal expansion at low temperatures of UO_2 and UO_2/ThO_2, J. Low Temp. Phys. 14:445 (1974).
9. G.K. White, Thermal expansion at low temperatures of the alkaline earth fluorides and PbF_2, J. Phys. C 13:4905 (1980).
10. S.I. Novikova, Investigation of thermal expansion of GaAs and ZnSe, Soviet Phys. - Sol. State 3:129 (1961).
11. G.K. White, Thermal expansion of cuprous oxide at low temperatures, J. Phys. C 11:2171 (1978).
12. G.K. White, Thermal expansion of $KTaO_3$ at low temperatures, J. Phys. C 14:L297 (1981).
13. B. Yates, R.F. Cooper and M.M. Kreitman, Low-temperature thermal expansion of zinc oxide, Phys. Rev. B 4:1314 (1971).
14. J.S. Browder, The thermal expansion of magnesium fluoride from room temperature to 16 K, J. Phys. Chem. Solids 36:193 (1975).
15. G.K. White, Thermal expansion of tellurium, J. Phys. C 6:1548 (1973).
16. A.C. Bailey and B. Yates, Anisotropic thermal expansion of pyrolytic graphite at low temperatures, J. Appl. Phys. 41:5088 (1970).

THE THERMAL EXPANSION COEFFICIENT OF FCC METALS

Rosemary A. MacDonald

Thermophysics Division
National Bureau of Standards
Washington, D.C.

INTRODUCTION

The work to be reported here improves upon and extends that presented at the 7th European Thermophysical Properties Conference in 1980[1]. There it was shown that the calculated thermal expansion values, ε, were very sensitive to the curvature of the potential, $\phi(r)$, assumed for the nearest neighbor interaction. With the modified Morse potential used in the present work, we obtain excellent agreement between ε and experiment. The other properties that can be obtained from the Helmholtz free energy, $F(V,T)$, using the appropriate thermodynamic relations, have also been calculated, i.e., the specific heats at constant volume, C_v, and at constant pressure, C_p, the coefficient of linear expansion, α, and the isothermal and adiabatic bulk moduli, B_T and B_s, respectively. The calculations were carried out for the fcc metals Cu, Ag, Ca, Sr, $A\ell$, Pb, and Ni. A detailed report of these results has been given in Ref. 2. In the present work we shall focus on the coefficient of linear expansion for which, in most cases, the calculated values are in surprisingly poor agreement with experiment, considering the results for ε. In view of the importance of α in thermodynamic relationships, such as the $C_p - C_v$ correction and the Gruneisen constant, it seemed worthwhile to investigate the reason for this unexpected discrepancy. For this purpose, the calculations were carried out for copper. Although the discrepancy is quite small in this case, copper is one metal for which we can compare two different methods of deriving α. In one[3,4], $\Delta L/\Delta T$ is obtained directly from the measured values of the specimen length, L, and a correction is made for the finite interval ΔT. In the other method, ε $(=(L - L_o)/L_o)$ is obtained directly from the data, and α is derived by differentiation of $\varepsilon(T)$ with respect to temperature. In the widely-used compilation, the TPRC-Data Series[5], a polynomial representation of $\varepsilon(T)$ is given for this purpose (and to interpolate values of ε).

First we give a brief summary of the theory pertinent to our calculation of the thermodynamic properties of the fcc metals. Next we present the results that were obtained for the thermal expansion of copper and then we examine the problem of obtaining α by derivation from thermal expansion data. We find that there can be quite a large variation in the values of α derived from thermal expansion data and we urge that the "direct" method of measurement[4] of α be used in future.

THEORY

Free Energy

The Helmholtz free energy, $F(V,T)$, expressed as the sum of the static lattice energy (F_o) and the harmonic (F_h) and lowest order anharmonic (F_A) terms of lattice dynamical perturbation theory, can be calculated for a nearest-neighbor central-force model of a monatomic fcc crystal in the high temperature limit ($T>\Theta_D$, the Debye temperature). The various contributions to the free energy are as follows,

$$F(r,T) = F_o + F_h + F_A \tag{1}$$

where, in the notation of Ref. 2,

$$F_o = 6N\,\phi(r),$$

$$F_h = 3N\left[\frac{F_1(r)}{kT} + F_3(r)kT - kT\ln kT\right],$$

$$F_A = 3N\left[F_2(r) - 2\phi(r) + F_4(r)\,(kT)^2\right].$$

The volume is represented by r, the nearest-neighbor separation.

It is crucial to the success of these calculations that the free energy can be determined as a function of volume, and this has been made possible by Shukla's calculation[6] of the Brillouin zone sums that enter the functions $F_i(r)$ (see Ref. 1 for details). These sums depend on volume implicitly through the parameter, a_1, defined by

$$a_1 = \frac{\phi'(r)/r}{(\phi''(r) - (\phi'(r)/r))}, \tag{2}$$

where $\phi'(r)$ and $\phi''(r)$, the first and second derivatives of the pair potential with respect to r, can be evaluated[1,6] at any value of a_1 in the range $-0.1 \leq a_1 \leq 0.1$, the range of practical interest. It remains, then, to define the pair potential.

Modified Morse Potential

As mentioned in the introduction, our preliminary calculations of ε with a Morse potential[1], indicated that a change in the curvature of the potential at interatomic separations other than r_o, the potential minimum, might bring ε into agreement with experiment. To this end, we have modified the Morse potential in the following way:

$$\phi(r) = \frac{D_o}{(1-2b)}\left[\exp(-2a(r-r_o)\sqrt{b}) - 2b\exp(-a(r-r_o)/\sqrt{b}\right] \quad (3)$$

The usual form is obtained when b=1. The additional parameter, b, affects the curvature away from but not at the minimum of the potential well. Its value is determined during the course of the calculation by requiring that ε fit the experimental results in the neighborhood of Θ_D. The values of the other parameters, D, a, and r_o are not affected by the introduction of b since they are determined at absolute zero, where it is assumed that all atoms are at the minimum of their potential wells.

Thermodynamic Properties

The equilibrium nearest-neighbor separation, r(T), is determined from the minimization equation,

$$F'(r,T) = 0, \quad (4)$$

where the prime denotes differentation with respect to r at constant temperature. The thermal expansion is then obtained from the definition

$$\varepsilon(T) \equiv \frac{r(T) - r(293)}{r(293)} \quad . \quad (5)$$

We have chosen 293 K as the reference temperature since it is the one most commonly used in experimental work. The coefficient of linear expansion, to which we devote the rest of this paper, is given by the relation

$$\alpha(T) = \frac{1}{r(T)}\left(\frac{dr}{dT}\right)_P \quad , \quad (6)$$

(note that r(293), instead of r(T), appears in the definition of α used in Refs. 4 and 5). The other thermodynamic properties involve derivatives of F(r,T) with respect to r and T, e.g., the specific heat at constant volume,

$$C_v(T) = -\,T \left(\frac{\partial^2 F(r,T)}{\partial T^2}\right)_{r(T)} \quad , \qquad (7)$$

and the isothermal bulk modulus,

$$B_T(T) = \frac{\sqrt{2}}{9\,r(T)}\,F''(r,T) \quad , \qquad (8)$$

where $F''(r,T)$ is the second derivative of $F(r,T)$ with respect to r at constant temperature. These properties have been discussed in Ref. 2. The derivatives F', F'' and dr/dT were all obtained by using a five point numerical differentiation formula[7] with an interval of .001. A smaller interval did not alter the results.

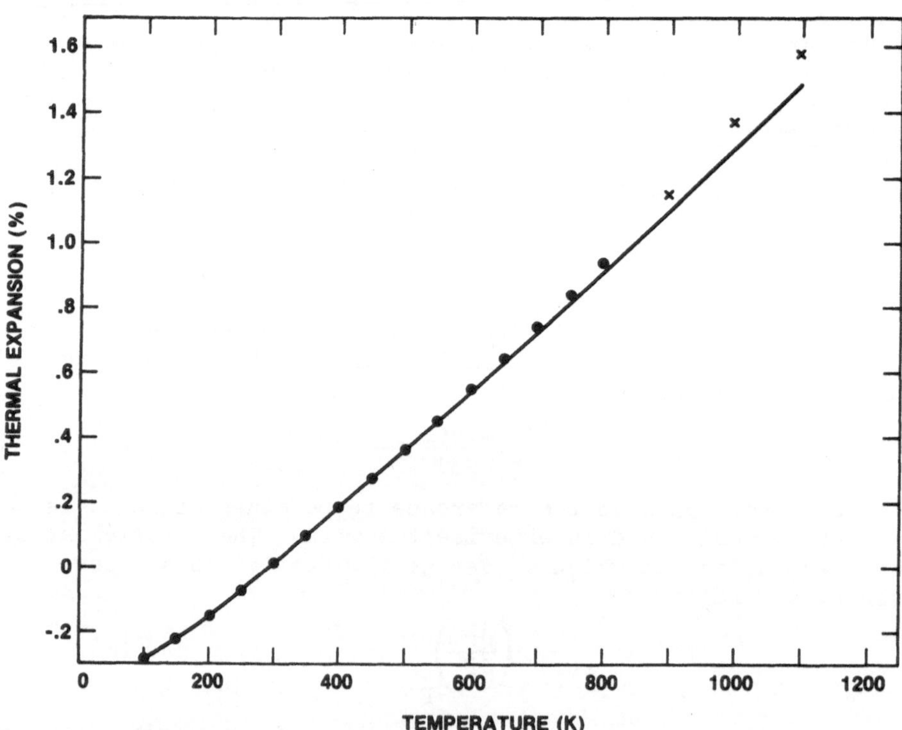

Fig. 1. Thermal expansion (%) of copper. Present calculation: εth, ——— ; experiment: ●[4], x[5].

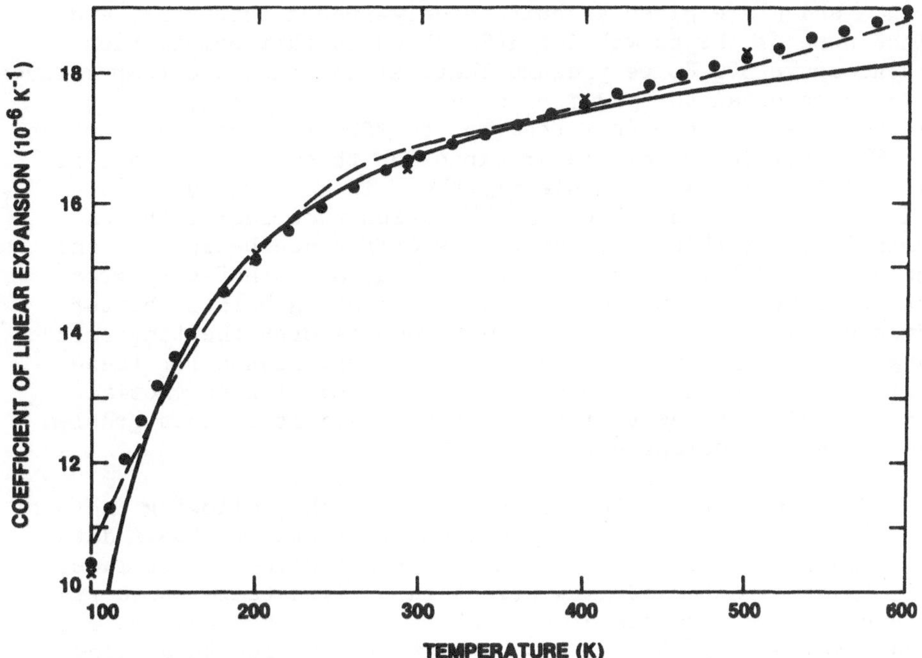

Fig. 2 Coefficient of linear expansion of copper. Present calculation: α^{th}, ———— ; derived from TPRC-polynomial representation[5] of thermal expansion: α_p,— — —; TPRC-recommended values of coefficient of linear expansion: α^{TPRC}, x; "direct" method of determining α from experimental data[4]:●

RESULTS

Here, we present the results of our calculations for copper. Figures 1 and 2 show the thermal expansion and the coefficient of linear expansion, respectively. For the thermal expansion there is excellent agreement with experimental data over quite a large range of temperature. However, as we noted in the introduction, α is not in as good agreement as we would have expected, judging by the results for ε. Since $r(T)$ can be computed over as finely divided an interval as is necessary to obtain accurate derivatives, we investigate the treatment of the experimental data to find the reason for this discrepancy. We use The TPRC Data Series[5] as a data source. The recommended values of α for copper, α^{TPRC}, are plotted in Fig. 2 for comparison with our calculated values, α^{th}. In this figure, we have restricted the temperature range to 100 — 600 K, the region where ε is in excellent agreement with experiment (see Fig. 1). Since the recommended values of both α and ε are given at quite widely-spaced intervals, and since none of the curves for $\alpha(T)$ shown in this compilation extend above 300 K, we presume that, at intermediate temperatures, α is to be obtained by differentiation of the polynomials provided[5], which have been fitted to the TPRC-recommended values of ε. The coefficient of linear expansion thus obtained, α_p, is also plotted in Fig. 2. Clearly there is a difference between α_p and the recommended values, α^{TPRC}, which underscores the unreliability of this procedure. The difference between α_p and the values of Ref. 4, also shown in Fig. 2, is not very great, but it is by no means negligible. In fact, α^{th} is in better agreement with these "direct" measurements over the temperature range 160 —— 360 K. One suspects that the reason for these differences lies in the problems associated with polynomial fitting and subsequent differentiation, and it is this problem that we shall address here.

We have examined the effect of α_p on the following factors:
(1) the order of the polynomial, m, fitted to the ε data,
(2) the range of temperature over which the fit is made,
(3) the number of data points in a given temperature range.
We used both our values of thermal expansion, ε^{th} and the recommended values[5], ε^{TPRC}, as sets of data points for this calculation. We shall refer to the coefficient of linear expansion obtained from ε^{th} as α_p^{th}, and that obtained from the ε^{TPRC} as α_p^{TPRC}. The results are as follows:

(1) Figure 3 shows the effect of changing the order of the polynomial from 3 to 6. We see that the polynomial fits to the TPRC data points, as demonstrated by the values of α_p^{TPRC}, are quite unstable. By contrast, the polynomial fits to the calculated values, and the values of α_p^{th}, are insensitive to the order of the polynomial.

(2) Figure 4 shows the effect of varying the range of temperature, T_1 —— T_2, used in the fitting procedure. Once again, there is a significant effect on α_p^{TPRC} when either T_1 or T_2 is changed, and, for the cases that we have considered, there is only fair agreement with α^{TPRC} (see upper set of curves). Again by contrast, there is good agreement between α_p^{th} and α^{th} when T_1 = 300 K, and, in this case, variation of T_2 has little effect. However, when T_1 is decreased, the polynomial fits also become sensitive to the temperature range and α_p^{th} is no longer in good agreement with α^{th} (see lower set of curves).

(3) The change in number of data points for a given temperature range appears to have no significant effect on either α_p^{th} or α_p^{TPRC}.

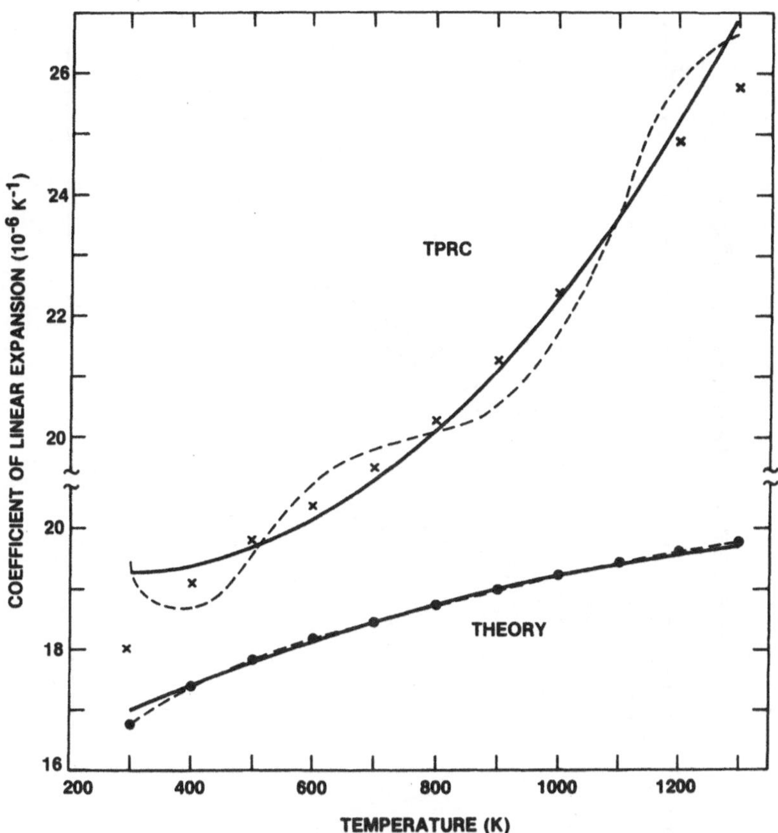

Fig. 3 Effect on coefficient of linear expansion of changing order of polynomial fitted to thermal expansion data for copper. Third order polynomial: ——— ; sixth order polynomial: — — — ; α^{th}: ● ; TPRC-recommended values: α^{TPRC}, x. Note break in ordinate in this figure and in Fig. 4, that separates the curves derived from ε^{TPRC} (upper set) from those derived from ε^{th} (lower set).

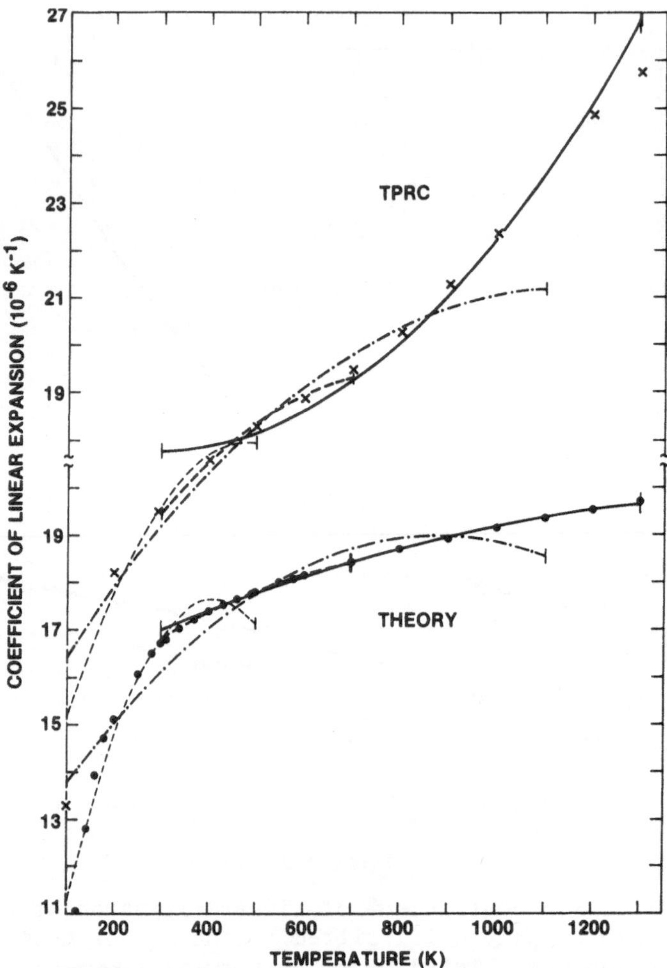

Fig. 4 Coefficient of linear expansion derived from polynomial fits of order 3 to thermal expansion data for copper over several temperature intervals. 300-1300 K, ———— ; 300-700 K, —— —— ——; 100-1100 K, —— · —— ; 100-500 K, - - - ; α^{th}, ● ; α^{TPRC}, x.

We note that all the above polynomials gave an acceptable fit to the thermal expansion data.

DISCUSSION

From the trials described above, we find that polynomial fits to the thermal expansion are generally sensitive to the temperature range over which the fit is made, insofar as the derivative of the polynomial is concerned. Further, even when the temperature range is favorable, the polynomial fits to experimental data are sensitive to the order of the polynomial, apparently as a result of the inherent scatter in the experimental points. What is particularly disturbing in these results is the fact that the method of obtaining the coefficient of linear expansion by differentiation does not give a satisfactory representation of either our results or the experimental values of α[4,5] over the range of temperature, 100-600 K, where the thermal expansion values are all in agreement. We can only conclude that the differentiation method of obtaining the coefficient of linear expansion is unreliable. Therefore, we urge that the method of measurement described in Ref. 4 be adopted as the preferred procedure in the future.

REFERENCES

1. R. C. Shukla and R. A. MacDonald, Thermal Expansion of Cubic Crystals, High Temperatures-High Pressures, 12: 291 (1980).
2. R. A. MacDonald and W. M. MacDonald, Thermodynamic Properties of FCC Metals at High Temperatures, Phys.Rev.B. 24:1715 (1981).
3. T. A. Hahn, Thermal Expansion of Cu from 20 to 800 K-Standard Reference Material 736, J.Appl.Phys., 41: 5096 (1970).
4. R. K. Kirby and T. A. Hahn, Nat.Bur.Stand. Certificate, Standard Reference Material 736, Washington, D.C., (1975).
5. Y. S. Touloukian, R. K. Kirby, R. E. Taylor, and P. D. Desai, in "Thermal Expansion, Metallic Elements and Alloys", Vol.12 of "Thermophysical Properties of Matter" (The TPRC Data Series), Touloukian Y. S., Ho, C. Y., eds. Plenum Press, New York (1975).
6. R. C. Shukla, On the Anharmonic Contribution to the Specific Heat of Monatomic FCC crystals, Int. J. Thermophysics, 1: 73 (1980).
7. P. J. Davis and I. Polonsky, in "Handbook of Mathematical Functions", M. Abramowitz and I. A. Stegun, eds., NBS Applied Mathematics Series 55, U. S. National Bureau of Standards, Washington, D. C. (1964), p. 914.
8. Y. A. Chang and R. Hultgren, The Dilation Contribution to the Heat Capacity of Copper and α-Brass at Elevated Temperatures, J. Phys. Chem., 69: 4162 (1965).
9 R. Hultgren, P. D. Desai, D. T. Hawkins, M. Gleiser, K. K. Kelly, and D. D. Wagman, in "Selected Values of Thermodynamic Properties of Metals and Alloys", American Society for Metals, Cleveland, Ohio (1973).

THE ROLE OF THERMAL EXPANSION IN THE MELTING AND SUPERIONICITY OF ALKALI AND ALKALINE EARTH HALIDES

L.L. Boyer

Naval Research Laboratory
Washington, DC 20375

Results of ab initio equation-of-state calculations for alkali and alkaline earth halides imply that melting and superionicity in these materials are related to different types of instabilities of the perfect crystalline solid. The instability associated with melting is the vanishing of the isothermal bulk modulus, brought on by the decrease in a shear elastic constant with increasing volume. Superionicity in the alkaline earth halides appears to be related to the softening of a zone boundary phonon with X_3 symmetry. The theory, as it relates to melting in alkali halides, is supported in a large measure by thermal expansion data. Good agreement between theory and experiment is also obtained for thermal expansion in the alkaline earth halides, but more high temperature data are needed. The theory predicts a relatively larger thermal expansion should accompany the transition to the superionic phase in SrF_2 than in CaF_2.

I. INTRODUCTION

A relatively simple first-principles (parameter-free) theory has been employed recently in equation-of-state calculations for alkali halides,[1-4] alkaline earth halides,[4,5] fluoperovskites,[6] and alkaline earth oxides.[7] In this theory the method of Gordon and Kim[8,9] is adopted for calculating pair potentials from ionic charge densities; then these pair potentials are used, in the quasiharmonic approximation, to derive the equation-of-state.

The results of these calculations have two important implications. First of all, they demonstrate that reasonably accurate equations of state, at nonzero temperatures, can be derived from first-principles. (Other workers[10,11] have applied the Gordon-Kim method in $T = 0$, *static* energy calculations, for a wide variety of ionic compounds. First-principles $T = 0$ equation-of-state calculations have also been performed for metals and semiconductors by other methods.[12,13]) Secondly, the results show a correlation between the various phase transitions in these materials (alkali halides, alkaline earth halides, and fluoperovskites) and predicted instabilities of their crystalline phases.

To illustrate the second point, consider specifically the compounds RbF, CaF_2, and $RbCaF_3$, formed by bringing together the proper proportion of Rb^+, Ca^{++}, and F^-

ions. Using Hartree-Fock free ion charge densities for these ions, and the Gordon-Kim prescription, pair potentials for the six possible ion pairs are readily calculated. These pair potentials, $\phi_{AB}(r)$, approximate the total energy of the A and B ions, separated by a distance r, minus the total energies of the constituent ions if they were isolated from each other.

The ϕ_{RbRb}, ϕ_{FF}, and ϕ_{RbF} potentials yield an equation of state for RbF which is in good agreement with experiment. At a temperature near the measured melting temperature an instability is predicted which is consistent with the observed properties of the melting transition. This instability was originally associated with melting in rare gas solids by Herzfeld and Goeppert-Mayer[14] and Kane.[15] It is characterized by the vanishing of the isothermal bulk modulus, which is brought on, in the alkali halides, by the decrease in the shear elastic constant, $C_{11} - C_{12}$, with increasing volume.

When the ϕ_{CaCa}, ϕ_{FF}, and ϕ_{CaF} potentials are used to calculate an equation of state for fluorite, one again finds good agreement with experiment. For this material an instability arising from vibrational modes with X_3 symmetry is predicted to occur before the crystal melts. This prediction correlates with the fact that CaF_2 transforms to a superionic solid phase with a disordered fluorine sublattice (consistent with X_3 symmetry) at a temperature below its melting temperature.

Finally, when all six potentials (ϕ_{RbRb}, ϕ_{CaCa}, ϕ_{FF}, ϕ_{RbF}, ϕ_{CaF}, and ϕ_{RbCa}) are used to calculate an equation of state for $RbCaF_3$, one again finds good agreement with experiment. According to the theory, the system prefers the perovskite structure for $T \geq 125$ K, even though it is metastable with respect to coordinated rotation (R_{25} symmetry) of the CaF_6 octahedra. Below ~ 125 K the free energy possesses a double well minimum as a function of the rotation angle, indicating a transition to a lower symmetry phase where these rotations are "frozen" into the structure. This type of phase transition is observed to occur in $RbCaF_3$ at ~ 196 K.[16] The physical origin of the preference for the distorted structure at low temperatures is simply that it is the more efficient way of packing the ions given the particular pair potentials involved. Specifically, in the perovskite structure, a large repulsive force between the Ca^{++} and F^- ions begins to develop as the volume decreases. But, a small rotation of the CaF_6 octahedra will permit a further decrease in volume, with the associated gain in Madelung energy, without further shortening the Ca–F bonds. The rotation is balanced by increased Rb – F repulsive forces due to reduced Rb – F separations that accompany the distortion.

Collectively, these results provide a unified picture for three different types of phase transitions (melting, in alkali halides; superionicity in alkaline earth halides; and displacive transitions, in the fluoperovskites) by a correlation, in each case, with an instability in the lower temperature phase. The fact that this is accomplished without input from experiment (other than universal constants) gives the correlation a large measure of credibility.

In the quasiharmonic approximation, the fundamental vibrational excitations are assumed to depend only on the variables that specify the crystal structure. In fact, of course, they also have an explicit dependence on temperature. But, the fact that the quasiharmonic theory successfully predicts these various crystal instabilities and their associated phase transitions shows that the primary effect of temperature in producing the instabilities is simply to thermally expand the structure to the region where the instability is present. It is therefore meaningful to compare the predictions of the theory for thermal expansion in the vicinity of the instability with the corresponding experimental results near the phase change.

An analysis of thermal expansion in the context of melting in the alkali halides was reported earlier.[2,3] The results for the alkali halides predict that the isothermal bulk modulus, β_T, goes to zero at a temperature, T_c, near the melting temperature, T_m. (For the alkaline earth halides we shall see that both the $\beta_T \to 0$ instability and an instability against X_3 symmetry vibrations are important.) As $T \to T_c$, the lattice constant may be written,

$$a(T) = a_0 + a_1 (T_c - T)^{1/2}, \quad (1)$$

which produces a divergent thermal expansion coefficient, $\alpha \propto (T_c - T)^{-1/2}$, as $T \to T_c$. On the whole, the thermal expansion data for the alkali halides tend to support these ideas, but more accurate data are needed near the melting temperature. Specifically, most of the data show a greater-than-linear increase in α as $T \to T_m$, which could be interpreted as the onset of the $\beta_T \to 0$ instability at $T_c \geqslant T_m$. For example, when Eq. (1) is used to fit data for NaCl at high temperatures, with a_0, a_1, and T_c treated as adjustable parameters, one obtains an excellent fit, but the optimum value of T_c can vary from T_m to $\sim 1.1\ T_m$ depending on the source of the data. These results are discussed in detail in Ref. 2.

The vanishing of the isothermal bulk modulus, in the alkali halides, is produced by the decrease in the quasiharmonic shear elastic constant, $C_{11} - C_{12}$, with increasing volume. It is natural to associate the volume where $C_{11} - C_{12} = 0$ with the volume of the liquid[17,18] since, in either case, there would be no resistance to shear stress. With this association one finds that not only does the theory give a good prediction for the thermal expansion of the solid, but in addition, the volume discontinuity that occurs upon melting is also accurately predicted.[4]

The stability of the fluorite structured alkaline earth halides, at high temperatures (near melting), is complicated, at least in theory, by the presence of another type of instability; specifically, that due to vibrations with wave vector at the (100) zone boundary having X_3 symmetry. It is the presence of this X_3 symmetry instability that appears to be the fundamental origin of superionicity in these materials.[4,5] For CaF_2 the quasiharmonic frequency of the X_3 mode, ν_{X_3}, is predicted to go to zero at a temperature (volume) that is near but smaller than the temperature (volume) where $\beta_T \to 0$. However, results for SrF_2 (below) have ν_{X_3} going to zero at a greater volume than does β_T, but still at a lesser volume than the shear instability (in this case $C_{44} \to 0$) which produces the $\beta_T \to 0$ instability. From the result that $\nu_{X_3} = 0$ at a smaller volume than $C_{44} = 0$, one still expects to find a superionic phase before the liquid phase in SrF_2 just as in CaF_2; indeed, this is born out experimentally. However, because of the different order of occurrence of the $\beta_T = 0$ and $\nu_{X_3} = 0$ instabilities in the two compounds, one would expect a qualitatively different result for thermal expansion at the superionic transition. In particular, a relatively larger thermal expansion would be expected to accompany the transition to the superionic phase in SrF_2 than in CaF_2.

In section II, the approximations of the theory are reviewed and discussed. Results for the vibrational frequencies and equation of state, of SrF_2, along with thermal expansion results for SrF_2 and CaF_2 are presented and discussed in section III.

II. APPROXIMATIONS OF THE THEORY

First of all, one needs an ab initio method for determining the charge density of the ions in question. For some compounds, e.g., fluorides, the charge densities of the free ions are adequate. These can be obtained from published tables[19] of Hartree Fock wave functions. For other compounds, such as oxides, it is necessary to include the effect of the self-consistent crystal field on the ionic density.[7,11]

Given the charge densities of the ions, the Gordon-Kim method is applied to calculate pair potentials for the various ion pairs. This involves two approximations:

(a) The charge density of an ion pair is taken to be the sum of the charge densities of the separate ions. In other words, the charge density of one ion is assumed not to be distorted by the presence of the other ion. This is the rigid-ion approximation.

(b) Next, the electronic ground state energy is derived from the charge density as though it were, locally, a free electron gas. This provides simple expressions for relating the kinetic, exchange, and correlation energies of the electrons to the charge density. The electrostatic contribution to the pair potential is derived from a straight forward, but somewhat tedious, application of Coulomb's law. This electron gas approximation is in the spirit of the Hohenberg-Kohn[20] theorem, which states that the ground state energy of any system of electrons is given by a universal functional of the charge density. The electron gas approximation therefore assumes that this universal functional, which is not known exactly, can be approximated by what is known to hold for the free electron gas.

Implicit in this derivation of the pair potential is the adiabatic approximation, which assumes the electrons stay in their ground state as the nuclei move. This is certainly valid for ionic compounds, even at temperatures well above melting.

Once the pair potentials are determined we approximate the electronic ground state energy, or potential energy, of the crystal as a sum over all the pair potentials; that is to say we make the pair potential approximation.

Given the pair potentials it is a straight forward matter to calculate the normal mode frequencies, ν_i, as a function of the volume, V, using the well known theory of lattice dynamics in the harmonic approximation.[21] In general, V denotes a set of variables which specify the positions of the ions in the chosen crystal structure. The fluorite structure requires only one such variable, namely, the volume. The $\nu_i(V)$ are precisely, the classical frequencies of small amplitude oscillations about the structure given by V. If we assume that the vibrational energy levels of the system are those of independent harmonic oscillators with frequencies $\nu_i(V)$, the free energy takes the simple form (Sec. IV of Ref. 21)

$$F(V,T) = U(V) + \frac{1}{2}\sum_i h\,\nu_i(V) + kT\sum_i \ln\{1 - \exp[-h\,\nu_i(V)/kT]\} \quad (2)$$

where U is the static contribution to the free energy, i.e., the sum of all pair potentials evaluated at the undistorted perfect lattice positions, T is the temperature, h is Planck's constant, and k is Boltzmann's constant. This assumption is known as the quasiharmonic approximation. One further approximation, the perfect crystal approximation, is implicit in the theory. The solid is assumed to be an infinitely extended perfect crystal with no allowance for thermal generation of defects, such as, vacancies or dislocations.

In Ref. 3 an attempt was made to assess the accuracy of these various approximations for the alkali halides by looking for a consistent explanation for the discrepancies between theory and experiment. It was noted that considerably too low values for thermal expansion were obtained when the difference between the size of the alkali and halide ions was large. It was suggested that a likely explanation for this trend could be the breakdown of the pair potential approximation. However, another factor in this trend could be the need for including self-consistent crystal field effects on the anion charge densities. (This is absolutely necessary for oxides since without the crystal field the O^{--} ion is unstable, while in the crystal, the bonding can be totally ionic.[7]) In either case, these errors are expected to be least for the fluorides.

Another important trend was noted between the accuracy of the thermal expansion results and the proximity of T_c to T_m. It was found that, in general, a $T_c > T_m$ ($T_c < T_m$) result was accompanied by a too low (high) value for room temperature thermal expansion. Furthermore, this trend could be made precise by assuming an anharmonic correction to T_c which would increase it by ~20%. If the pair potentials were perfectly accurate and the pair potential approximation was valid, then the quasiharmonic theory would produce accurate thermal expansion results at low temperatures. But, the quasiharmonic T_c would be, according to the trend, lower than T_m by ~20%. Consequently, the quasiharmonic result for α would steadily become much too high, with increasing temperature, owing to its divergent behavior at $T_c \sim 0.8\ T_m$ (see Eq. (1)). This empirically deduced lowering of α (from the quasiharmonic result), with increasing temperature, is consistent with the results of Feldman et al.[22] and Hardy,[23] who show that anharmonic corrections lower the effective Grüneisen parameter.

III. RESULTS

Charge densities for the Sr^{++} and F^- ions were taken from tables[19] of Hartree-Fock wave functions for use in calculating pair potentials for the SrSr, FF, and SrF interactions. The charge density for Sr^{++} was obtained by omitting the 5s electrons from the Sr atom; the tables in Ref. 19 do not include wave functions for Sr^{++}. This procedure was tested in a calculation on CaF_2, for comparison with results of earlier calculations[4] which used the true Ca^{++} density, and found to have very little effect on the results. However, the same procedure, if employed for anions would probably introduce large errors.

Values for the various contributions (electrostatic, kinetic, exchange, and correlation) to the pair potentials were calculated, using program POTLSURF,[24] at several separations in the range of interest for SrF_2. The kinetic, exchange, and correlation contributions, and the short range part of the electrostatic term were each fit to exponentials, $Ce^{-\beta r}$. The values for C and β are listed in Table I. The short range contribution to the SrSr potential was found to be negligible.

Calculated normal mode frequencies for the predicted room temperature volume are listed in Table II for comparison with experimental values. We see the agreement is quite good with the poorest results coming from the highest frequency modes. These are from the longitudinal optic branch and are known to be most affected by polarization. However, they do not contribute significantly to the equation of state since they are only weakly volume dependent.

TABLE I. Values obtained for C and β by fitting the exponential form, $Ce^{-\beta r}$, to calculated values of the short range Coulomb (SRC), kinetic (KE), exchange (EX), and correlation (CORR) contributions to the indicated pair potentials in the range r_l to r_u. Results are in atomic units with energy in Hartrees.

Ion Pair	C,β	SRC	KE	EX	CORR	r_l	r_u
SrF	C	−45.710	114.567	−11.620	−0.2701	4.2	5.4
SrF	β	1.7856	1.6861	1.3918	1.0513	4.2	5.4
FF	C	−17.798	25.746	−3.0525	−0.1172	5.2	6.4
FF	β	1.5567	1.4072	1.1115	0.8459	5.2	6.4

Table II. Comparison of calculated and experimental phonon frequencies (300 K) for SrF_2 at selected wave vectors in the Brillouin Zone.

Wave Vector	Method	Frequencies (cm^{-1})					
Γ	Calc.	226	311[b]	453			
Γ	Expt.[a]	213	280[b]	383			
X	Calc.	150	141[c]	173	217	350	429
X	Expt.[a]	137	153[c]	183	190	292	347
L	Calc.	110	201	262	272	338	367
L	Expt.[a]	130	187	245	250	280	307

a) Ref. 25
b) Raman active
c) X_3 symmetry

From the pair potentials and the normal mode frequencies, derived from the pair potentials, the free energy is easily determined numerically using Eq. (2). Details of this procedure are discussed in Ref. 3. The equation of state can be conveniently illustrated by plotting the static pressure, $P_s \equiv \partial U/\partial V$, and the vibrational pressure, $P_v(T) \equiv \partial(U - F)/\partial V$, for selected temperatures, as a function of lattice constant, a.

The calculated equation of state for SrF_2 is plotted in this way in Fig. 1. The value of a at temperature T is given by the intersection of P_s with $P_\nu(T)$. The calculated value of a at room temperature (5.747 Å) is ~1% smaller than the measured value (5.798 Å). Results for the thermal expansion of SrF_2 and CaF_2 are compared with experiment data in Figs. 2 and 3. An effective value of a for liquid SrF_2[33] (6.20 Å) agrees favorably with the calculated value for which $C_{44} = 0$ (6.27 Å).

Above a certain critical temperature, $T_c \sim 1770$ K the P_ν curves are larger than P_s at all volumes; at T_c the slopes of P_s and P_ν are identical and therefore, $\beta_{T_c} = 0$. The results show that $\beta_T \to 0$ at a smaller a than that for which $\nu_{x_3} = 0$: the reverse is true for CaF_2. Even though the $\beta_T \to 0$ instability occurs before $\nu_{x_3} \to 0$ in SrF_2, one still expects a superionic transition, rather than melting, because ν_{x_3} still vanishes at a smaller volume than does C_{44}. For SrF_2, $\nu_{x_3} \to 0$ appreciably nearer to the volume where $C_{44} = 0$ than occurs in CaF_2; since the P_ν curves diverge asymptotically at the volume where $C_{44} = 0$, this produces the qualitatively different result for the two compounds.

The predicted qualitative difference in the way CaF_2 and SrF_2 become unstable at high temperatures is consistent with the observed temperature dependence of the ionic

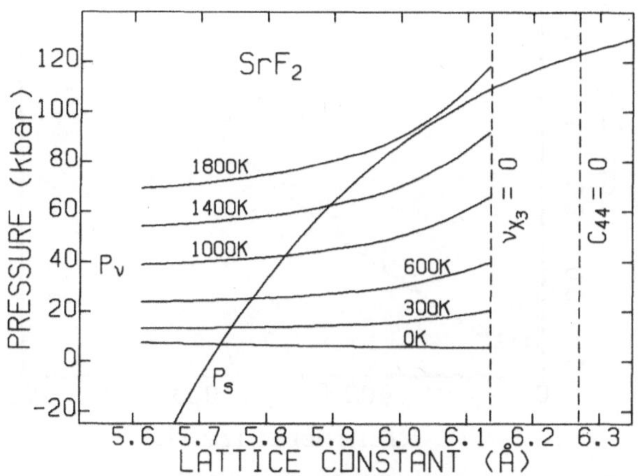

Fig. 1. Equation of State for SrF_2; plot of the static pressure (P_s) and vibrational pressure ($P_\nu(T)$), for selected temperatures, as a function of lattice constant

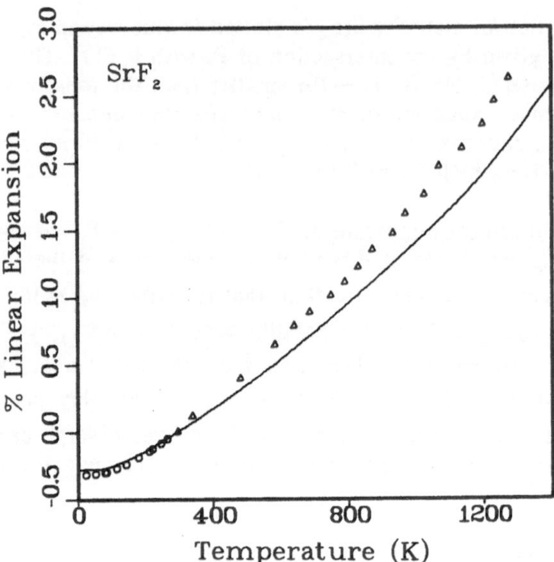

Fig. 2. Comparison of calculated thermal expansion for SrF_2 with experimental data (O—Ref. 26, Δ — ref. 27) as tabulated in Ref. 28

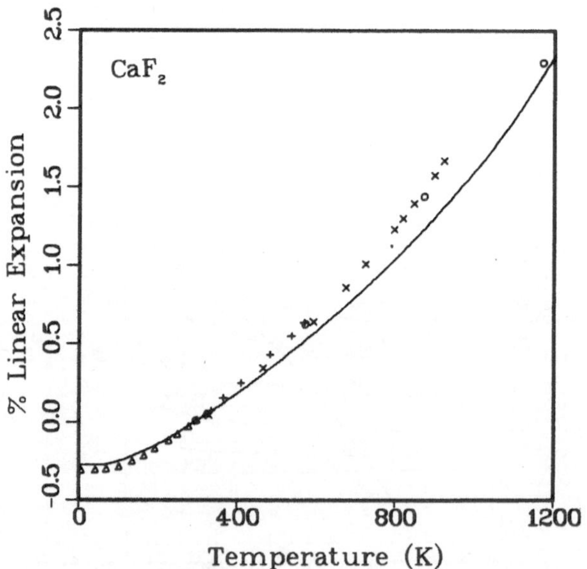

Fig. 3. Comparison of calculated thermal expansion for CaF_2 with experimental data (O — Ref. 29, Δ—Ref. 30, + —Ref 31, x —Ref. 32) as tabulated in Ref. 28.

conductivity in these materials. Specifically, SrF_2 has a relatively sharp increase in conductivity upon transformation to the superionic phase, while the transition in CaF_2 takes place more gradually.[34] The present theory suggests the sharp increase in conductivity for SrF_2 is caused by a correspondingly abrupt increase in volume. Unfortunately, experimental data are not available as sufficiently high temperatures in these compounds to test this hypothesis. However, larger-than-normal thermal expansion has been observed to accompany the superionic transitions in PbF_2 and $SrCl_2$,[35] and in BaF_2.[36] These results could indicate that superionic transitions in these materials are similar to that predicted for SrF_2; i.e., $\beta_T \rightarrow 0$ before $\nu_{x_3} \rightarrow 0$. In any case, it would be most desirable to have high temperature thermal expansion data for CaF_2 since it shows relatively little increase in conductivity at the superionic transition compared to other superionic materials with the fluorite structure.

REFERENCES

1. L. L. Boyer, Phys. Rev. Lett. 42, 584 (1979).
2. L. L. Boyer, "An Analysis of Thermal Expansion and Melting in Alkali Halides," Proceedings of the 7th International Thermal Expansion Symposium (Plenum, New York, to be published).
3. L. L. Boyer, Phys. Rev. B 23, 3673 (1981).
4. L. L. Boyer, Phys. Rev. Lett. 45, 1858 (1980); 46, 1172 (1981).
5. L. L. Boyer, "Origin of Superionicity in Alkaline Earth Halides," Proceedings of the International Conference on Fast Ionic Transport in Solids, May 1981, to be published in Solid State Ionics.
6. L. L. Boyer and J. R. Hardy, Phys. Rev. B, to be published.
7. L. L. Boyer, Bull. Am. Phys. Soc. 26, 391 (1981).
8. R. G. Gordon and Y. S. Kim, J. Chem. Phys. 56, 3122 (1972).
9. M. J. Clugston, Advan. in Phys. 27, 893 (1978).
10. A. J. Cohen and R. G. Gordon, Phys. Rev. B 12, 3228 (1975).
11. C. Muhlhausen and R. G. Gordon, Phys. Rev. B 23, 900 (1981).
12. V. L. Moruzzi, A. R. Williams, and J. F. Janak, Phys. Rev. B 15, 2454 (1977).
13. M. T. Yin and M. L. Cohen, Phys. Rev. Lett. 45, 1004 (1980).
14. K. F. Herzfeld and M. Goeppert-Mayer, Phys. Rev. 46, 995 (1934).
15. G. Kane, J. Chem. Phys. 7, 603 (1939).
16. H. Jex, J. Maetz, and M. Müllner, Phys. Rev. B 21, 1209 (1980).
17. M. Born, J. Chem. Phys. 7, 591 (1939).
18. J. L. Tallon, Philos. Mag. 39, 151 (1979).
19. E. Clementi and C. Roetti, Atomic Data and Nuclear Tables (Academic, New York, 1974).
20. P. Hohenberg and W. Kohn, Phys. Rev. 136, B864 (1964).
21. M. Born and K. Huang, Dynamical Theory of Crystal Lattices (Oxford University Press, London, 1954).
22. C. Feldman, J. L. Feldman, G. K. Horton, and M. L. Klein, Proc. Phys. Soc. London 90, 1182 (1967).
23. R. J. Hardy, J. Geophys. Research 85, 7011 (1980).
24. S. Green and R. G. Gordon, Quantum Chemistry Program Exchange, QCPE Program No. 251, Chemistry Dept., Indiana University.
25. M. M. Elcombe, J. Phys. C: Solid State Phys. 5, 2702 (1972).
26. A.C. Bailey and B. Yates, Proc. Phys. Soc. (London) 91, 390 (1967).

27. I. F. Ferguson, R. S. Street, and R. W. M. D'Eye, USAEC Report AERE-R-3344 (1960).
28. Thermal Expansion, Vol. 13 of Thermophysical Properties of Matter, edited by Y. S. Touloukian and C. Y. Ho (Plenum, New York, 1977).
29. O. J. Whittemore Jr. and N. N. Ault, J. Amer. Ceram. Soc. 39, 443 (1956).
30. D. N. Batchelder and R. O. Simmons, J. Chem. Phys. 41, 2324 (1964).
31. V. T. Deshpande and D. B. Sirdeshmukh, Ind. J. Pure Appl. Phys. 2, 405 (1964).
32. S. S. Sharma, Proc. Indian Acad. Sci. A 31, 261 (1950).
33. G. J. Janz, F. W. Dampier, G. R. Lakshminarayanan, P. K. Lorenz, and R. P. T. Tomkins, U.S. National Bureau of Standards, National Standards Reference Data Series — 15 (U.S. GPO, Washington, D.C., 1968).
34. C. E. Derrington, A. Lindner, and M. O'Keefe, J. of Solid State Chem. 15, 171 (1975).
35. M. H. Dickens, W. Hayes, and M. T. Hutchings, J. de Physique Colloque 37, C7-353 (1976).
36. D. O. Pederson, T. E. Duerr, R. B. Foster, and S. R. Montgomery, Bull. Am. Phys. Soc. 26, 404 (1981).

THERMAL EXPANSION OF TEN GRIMM-SOMMERFELD COMPOUNDS

Robert R. Reeber and John L. Haas, Jr.

U.S. Army Research Office
P.O. Box 12211
Research Triangle Park, NC 27709

U.S. Geological Survey
National Center, Stop 959
Reston, VA 22092

INTRODUCTION

Thermal expansion is an important property that relates directly to bonding forces between atoms in a solid material. Knowledge of thermal expansion also has practical value for the design of engineering devices and systems. Grimm-Sommerfeld compounds, those materials that have an average number of valence electrons per atom equal to four, are of interest because of their technical applications as semiconductors and geothermometers and additionally because of their importance as model materials for crystal-chemistry studies.

Although theoretical models of thermal expansion exist, they generally require a significant amount of information in addition to experimental measurements of length changes. On the other hand, most methods of fitting thermal-expansion data rely on empirical functions that have limited, if any, theoretical significance. Thus, comparison of experimental data from different sources often determined by differing techniques and from only partially overlapping data sets is difficult. A review of such fitting methods, appropriate for the needs of this work, has been given elsewhere.[1] Barron and coworkers[2] have also comprehensively reviewed theoretical approaches and the ranges of their agreement with experimental results.

Here we apply a method devised earlier[1] to an exhaustive set of available cryogenic and higher temperature thermal expansion and expansivity data reported in the literature for selected Group IV elements and a large number of important II-VI and III-V Grimm-

Sommerfeld compounds. This method allows the calculation of empirical parameters that can yield useful and physically quantitative information about solids. Such information results from the accurate (over extended temperature ranges) representations of thermal expansion/expansivity/temperature dependencies. Relying on a crude simplification of the lattice spectrum, the model can be considered that of a modified Einstein solid. The phenomenological approach, combined with an empirical least-squares trial-and-error deconvolution, provides a direct comparison of cryogenic with higher temperature thermal expansion and expansivity data. Estimates obtained by the method have been shown[1] to correlate with experimental results at temperatures significantly above the Debye temperature. Therefore, the method is useful for predicting reasonable values for higher temperature thermal expansions and molar volumes when higher temperature data are limited or non-existent.

CALCULATIONS

As has been discussed elsewhere,[1] the method relies on Blackman's less rigorous but more physically significant definition of Gruneisen's constant, γ_i[3]. Blackman defined it in terms of $\overline{\gamma}$, an arithmetic average of γ_i for a given interval, $\rho_\gamma(\nu_n)$, in the density of states. This more tractable expression for the Gruneisen constant may be approximated by a summation of three or more terms as follows:

$$\overline{\gamma} = \frac{\sum_{n=1}^{i} \overline{\gamma}_{\nu_n} \, \overline{\rho(\nu)}_n \, \overline{E}(h\nu_n/kT)}{\sum_{n=1}^{i} \overline{\rho(\nu)}_n \, \overline{E}(h\nu_n/kT)} \tag{1}$$

where

$$\overline{E}(h\nu_n/kT) = \frac{\left(\frac{\Theta_n}{T}\right)^2 \exp\left(\frac{\Theta_n}{T}\right)}{\left(\exp\left(\frac{\Theta_n}{T}\right)-1\right)^2} \tag{1a}$$

when θ_n are characteristic temperatures. Incorporating this expression for $\overline{\gamma}$ into Gruneisen's rule allows representation of the volume thermal expansion, α_v. Equations (2) and (3) are derived elsewhere.[1]

$$\alpha_V = \sum_{n=1}^{i} \overline{X}_n \overline{E}(h\nu_n/kT) \tag{2}$$

$$\ln \frac{V}{V_o} = [\sum_{n=i}^{i} \overline{Y}_n \overline{P}(h\nu_n/kT)]_0^T \tag{3}$$

Here, α_V is the volume thermal expansion, i is the number of Einstein frequencies, V is the molar volume, and $\overline{E}$ is the heat capacity associated with such averages. The coefficients $\overline{X}_n$ and $\overline{Y}_n$ are determined empirically by trial-and-error least-squares fitting of the empirical data, and

$$\overline{P}(h\nu_n/kT) = \frac{\Theta_n}{\exp[(\theta_n/T) - 1]} \tag{3a}$$

Theoretically, in the approximate lattice model, the coefficients $\overline{X}_n$ and $\overline{Y}_n$ include the compressibility, volume, Blackman's average $\overline{\gamma}_{\nu n}$ parameters and the heat capacity associated with these averages. Here the fitting procedure was simplified somewhat, in comparison to earlier work,[1] in that no attempt was made to fix the lowest temperature expansion with elastic parameters.

The convergence of the trial-and-error procedure is accomplished by minimizing the sum of the standard errors for the Einstein terms fitting the experimental data. The computer program utilizes a two-step procedure for optimizing the fitting. The program generates a table of 24 systematically different sets of frequencies, three frequencies in each set. After the best least-squares fit to experimental data is found from this table, a series of additional sets iterated from the best set is trial-and-error fitted until no further improvements are possible. Because most major features of the vibration distribution contribute to the low temperature data, only experimental data obtained at temperatures below or as high as approximately 100 kelvins above the Debye temperature θ_D are included in the calculations.

RESULTS

Table 1 lists quantitative fitting parameters for the thermal expansion of ten important Grimm-Sommerfeld compound semiconductors. Table 2 gives references to lattice-parameter and thermal-expansion work selected as most representative of each material. Selected

Table 1. Thermodynamic Fitting Parameters -- Characteristic Temperatures (Θ) Weighting Coefficients (W) -- and Lattice Parameters (a) and Debye Temperatures for Ten Grimm-Sommerfeld Compounds

Material	Θ_1	W_1	Θ_2	W_2	Θ_3	W_3	Cell Data a_o	Debye[a] Temp. Θ
	(K)	($X10^{-7}$ K^{-1})	(K)	($X10^{-7}$ K^{-1})	(K)	($X10^{-7}$ K^{-1})	(Å)	(K)
ZnS	10.625	-8.235	250.63	42.70	780	62.743	5.40947	349
ZnSe	80.	-12.13	294.375	82.07	780.62	16.21	5.66870	276
ZnTe	39.375	-7.90546	267.5	91.2971	701.875	10.6706	6.1026	222
CdTe	90.	-41.76	261.25	100.6	1136.25	2.448	6.48124	160
AlSb	23.75	-8.67406	477.5	72.1680	1103.75	-21.988	6.1355	294
GaAs	18.75	0.1166	96.875	-4.336	375.625	70.55	5.65315	346
GaSb	28.125	-2.233	111.25	-3.745	320.625	78.52	6.09575	267
InAs	75.	-15.9063	258.75	61.1426	1052.5	7.1227	6.0583	250
InSb	39.375	-17.8150	245.	71.9121	745.	3.9358	6.47877	207
HgTe	29.375	-47.3578	175.625	89.1259	517.5	11.2403	6.4576	138

[a]R. R. Reeber, The correspondence of lattice characteristic temperatures with Debye temperatures of some inorganic compounds, Phys. Stat. Sol. A 26:253 (1974).

Table 2. Selected References for Lattice Parameter and Thermal Expansion [RT, Room Temperature]

Material	Kelvin Temperature Range	Investigator	Date	Lattice Parameter (Temperature in Kelvins)	Ref.
ZnS	18-273	Adenstedt	1936		4
	284-1219	Skinner	1962	5.4093 (298)	5
	4.2-330	Reeber & McLachlan	1968	5.4095 (300)	6
	70-310	Browder & Ballard	1969		7
	3-32	Smith & White	1975		8
		Skinner et al.	1959	5.4094 (298)	9
		Boorman & Sutherland	1969	5.4098	10
ZnSe		Goryunova & Fedorova	1959	5.6686 (298)	11
	293-1004	Skinner	1962	5.6687 (293)	5
	4-85	Smith & White	1975		8
	313-523	Ballard et al.	1978		12
	299	Devlin et al.	1960	5.6687 (299)	13
ZnTe	20-340	Novikova & Abrikosov	1963		14
	295-310	Shiozawa et al.	1962		15
	296-731	Holland & Beck	1968	6.1026 (296)	16
	308-713	Singh & Dayal	1970	6.1026 (293)	17
	1.5-30	Collins et al.	1980		18
	299	Devlin et al.	1960	6.1037 (299)	13
	294	Reeber & Milton	1979	6.1028 (294)	19
CdTe	70-310	Browder & Ballard	1969		7
	10-300	Browder & Ballard	1972		20
	293-693	Williams et al.	1969	6.4809 (293)	21
	2-85	Smith & White	1975		8
	RT	Davis & Shilliday	1960	6.4815 (RT)	23
	300	Schaake	1969	6.4809 (300)	24
	293	Reeber & Milton	1979	6.4819 (293)	19
HgSe		Kleshchinskii et al.	1968	6.0854	22
HgTe	20-340	Novikova & Abrikosov	1963		14
	2-85	Collins et al.	1980		18
	55-300	Alper & Saunders	1967	6.461 (RT)	25
	RT	Lawson et al.	1959	6.460 (RT)	26
	RT	Parthe	1964	6.4623 (RT)	27
	RT	Reeber	1974	6.4577 (RT)	28
		Ivanov-Omskii et al.	1968	6.4588 (293)	29

Table 2. Continued

Material	Kelvin Temperature Range	Investigator	Date	Lattice Parameter (Temperature in Kelvins	Ref.
GaAs	296-1008	Shaw & Liu	1965	5.6528 (296)	30
	2-40	Sparks & Swenson	1967		31
	4-30	Smith & White	1975		8
	28-350	Novikova	1961		32
	288-338	Straumanis & Kim	1965	5.65321(298)	33
	211-473	Pierron et al.	1966	5.6532 (297)	34
	300-510	Sirota & Pashintsev	1958	5.6536 (301)	35
	78-673	Pashintsev & Sirota	1959	5.6536 (293)	36
	298	Ozolin'sh et al.	1963	5.65315(298)	44
	291	Giesecke & Pfister	1958	5.6534 (291)	38
GaSb	2-32	Sparks & Swenson	1967		31
	20-340	Novikova & Abrikosov	1963		14
	291	Giesecke & Pfister	1958	6.0954 (291)	38
	298	Ozolin'sh et al.	1963	6.09575 (298)	37
InSb	303-751	Shaw & Liu	1965	6.4785 (303)	30
	2-34	Sparks & Swenson	1967		31
	291	Giesecke & Pfister	1958	6.47877(291)	38
	30-340	Novikova	1960		39
	298	Ozolin'sh et al.	1963	6.47965(298)	37
	298	Swanson et al.	1955	6.4782 (298)	40
AlSb	293	Sirota & Gololobov	1962	6.136 (293)	41
	25-340	Novikova & Abrikosov	1963		14
	291	Giesecke & Pfister	1958	6.1355 (291)	38
		Parthe	1964	6.1361 (RT)	27
		Swanson et al.	1955	6.1347 (299)	42
InAs	78-673	Pashintsev & Sirota	1959	6.0572 (293)	36
	2-42	Sparks & Swenson	1967		31
	100-600	Sirota & Berger	1959		43
	292-528	Sirota & Pashintsev	1958	6.0572 (292)	35
	291	Giesecke & Pfister	1958	6.0584 (291)	38
	298	Ozolin'sh et al.	1963	6.05838(298)	44

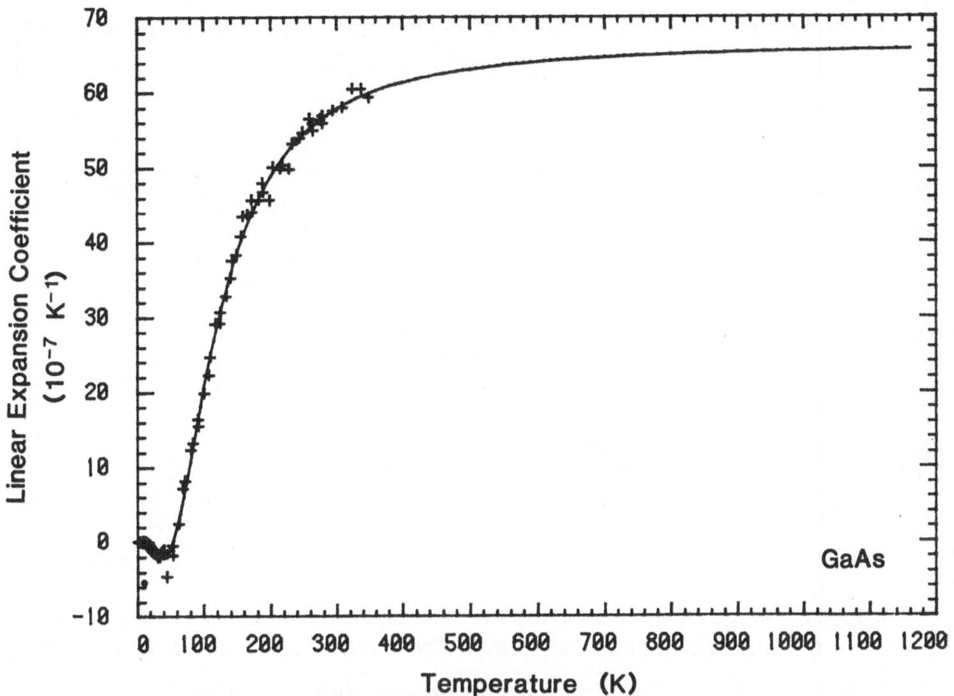

Figure 1a. Temperature dependence of thermal expansion for GaAs. Solid line is least-squares fit to the experimental data from sources listed in table 2.

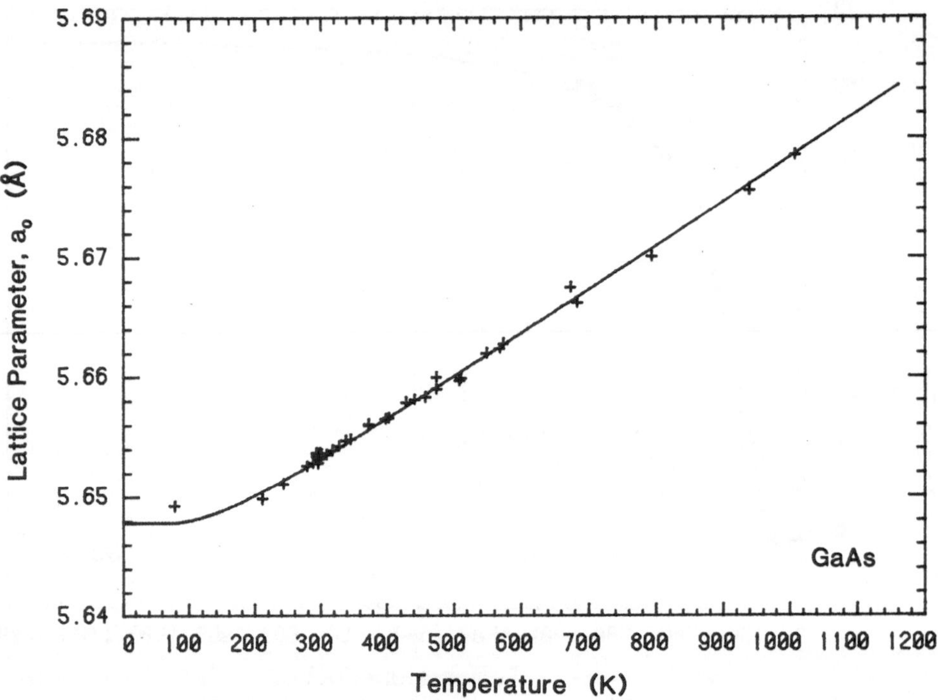

Figure 1b. Temperature dependence of lattice parameter for GaAs. Solid line is derived from least-squares fit shown in figure 1a.

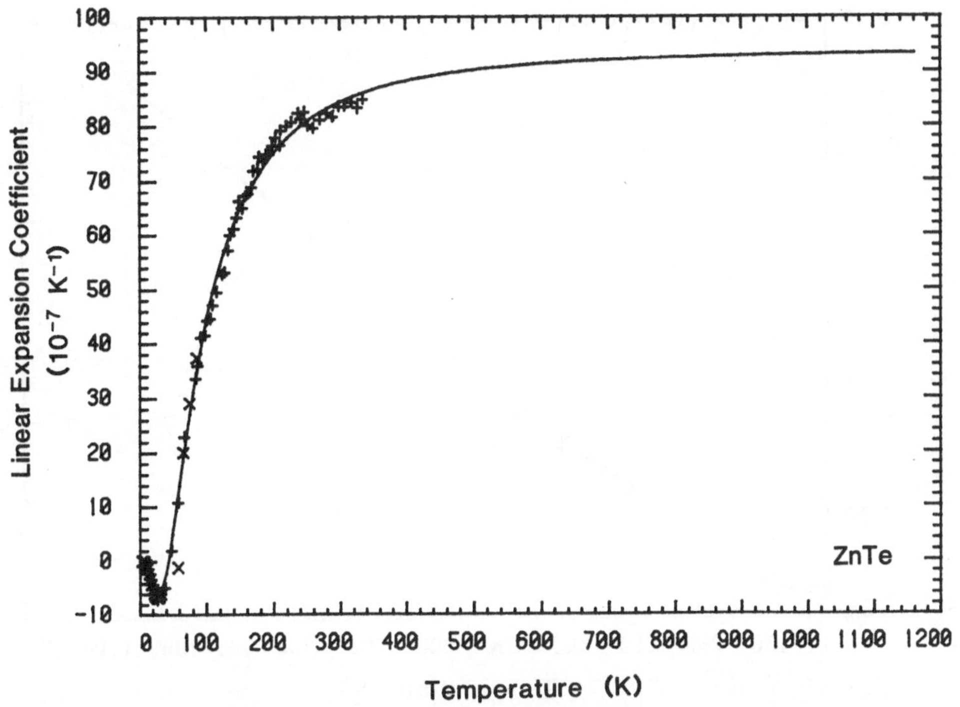

Figure 2a. Temperature dependence of thermal expansion for ZnTe. Solid line is least-squares fit to the experimental data from sources listed in table 2.

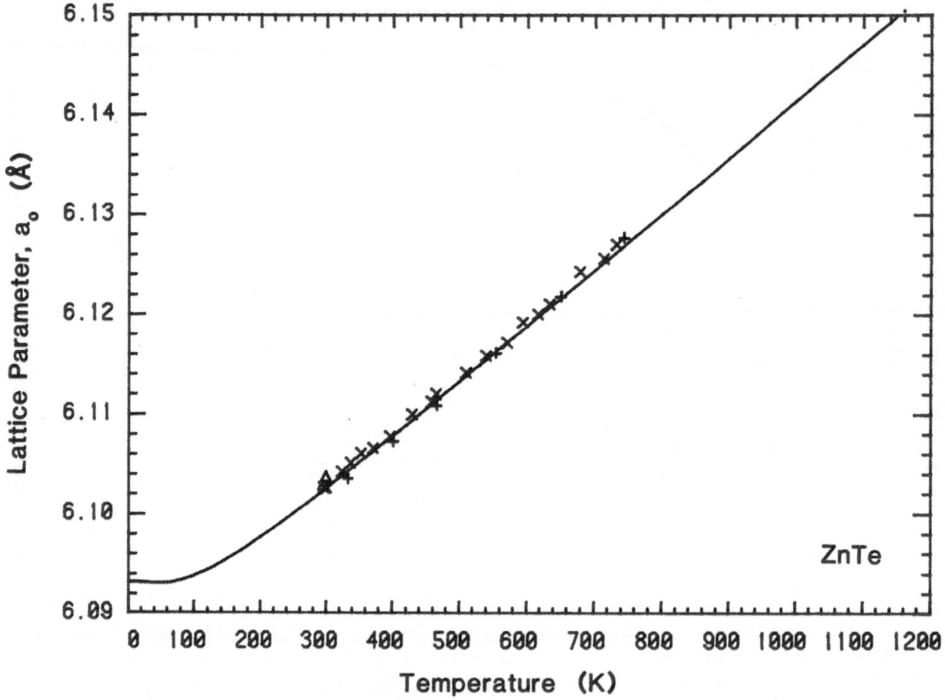

Figure 2b. Temperature dependence of lattice parameter for ZnTe. Solid line is derived from least-squares fit shown in figure 2a.

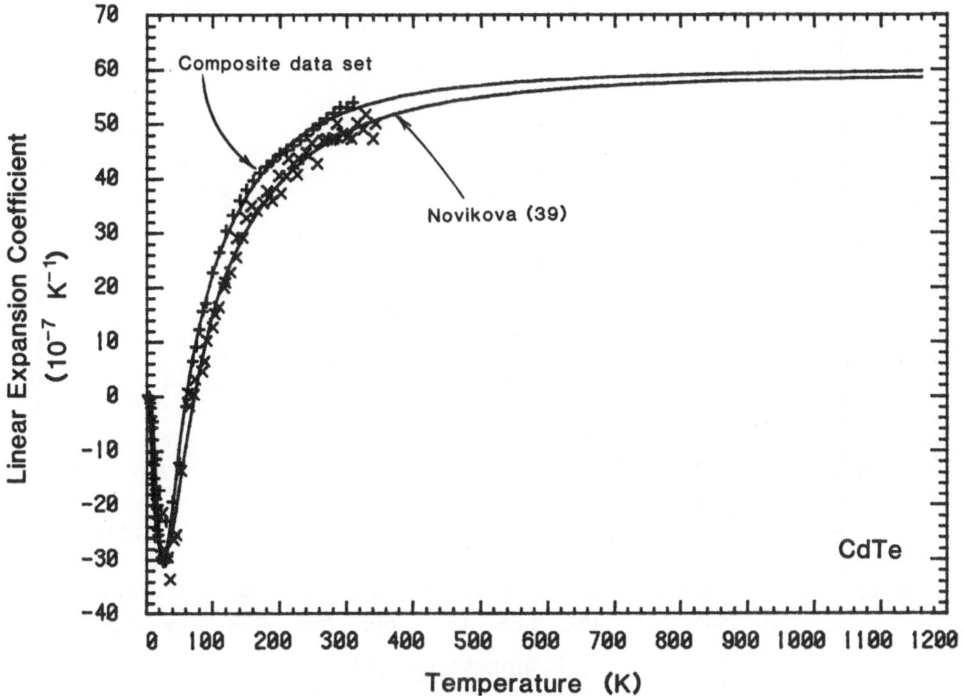

Figure 3a. Temperature dependence of thermal expansion for CdTe. Solid lines are least-squares fits of a composite data set from sources listed in table 2 (+) and of the data of Novikova[39] (x), as indicated.

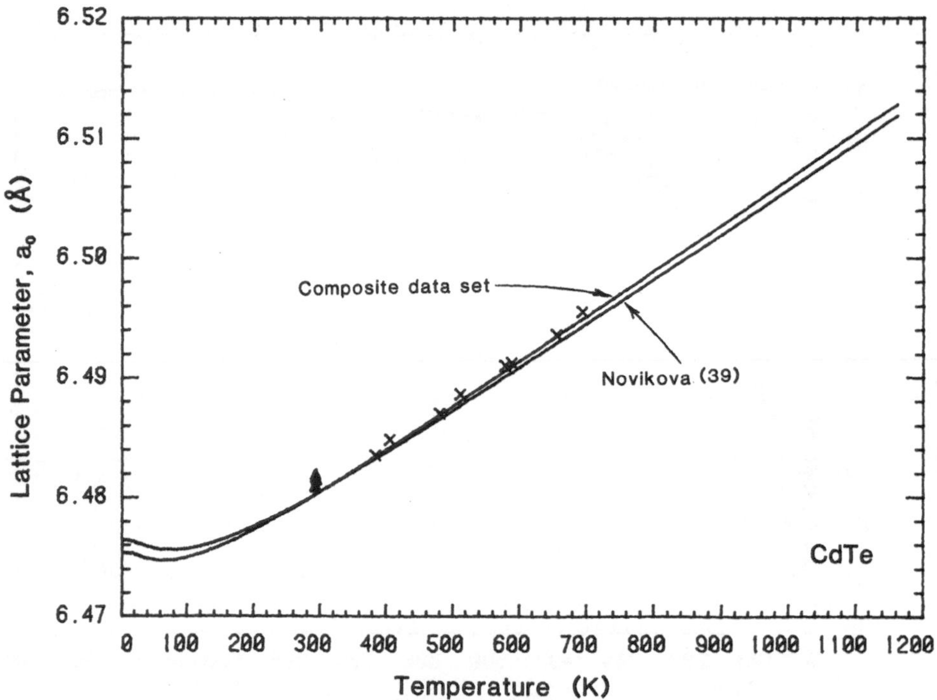

Figure 3b. Temperature dependence of lattice parameter for CdTe. Solid lines are derived from those shown in figure 3a. Of the two curves, that derived from the composite data set is in better agreement with the experimentally measured cell edges of Williams and coworkers[21].

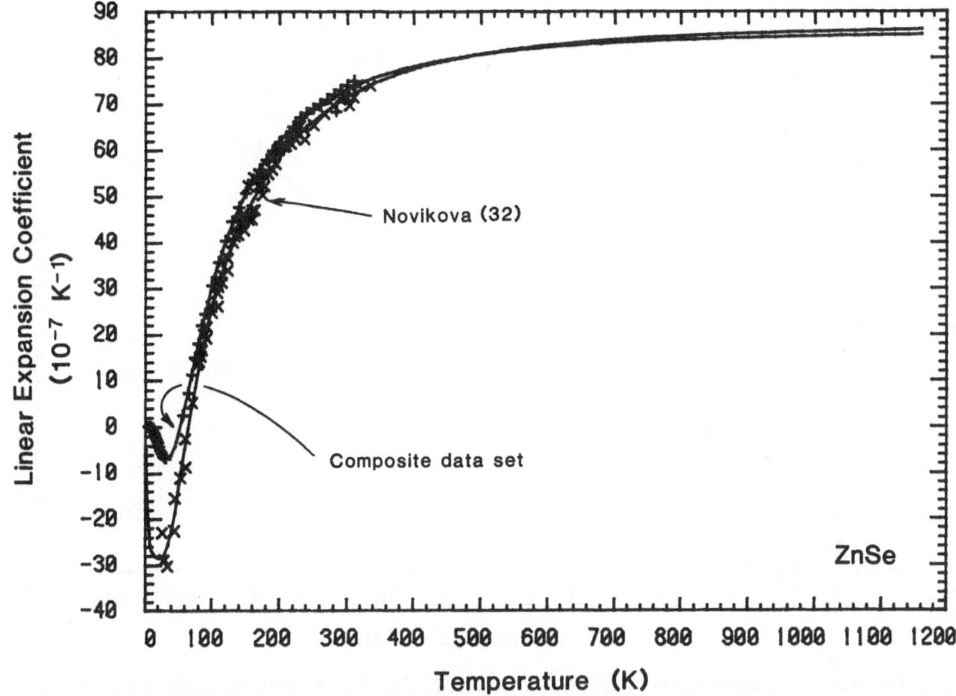

Figure 4a. Temperature dependence of thermal expansion for ZnSe. Solid lines are least-squares fits of a composite data set from sources listed in table 2 (+) and of the data of Novikova[32] (x), as indicated.

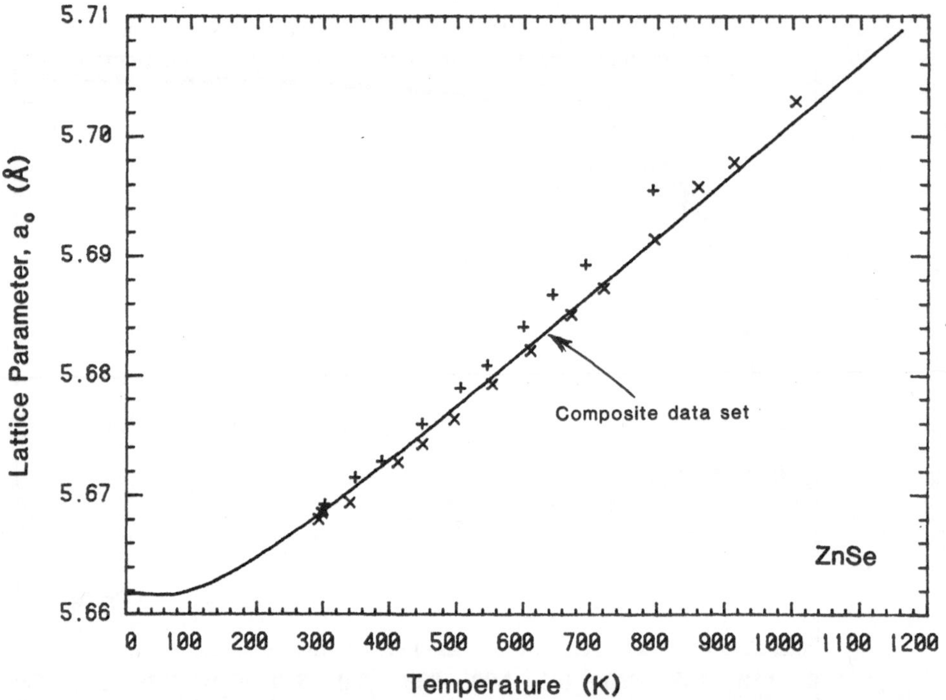

Figure 4b. Temperature dependence of lattice parameter for ZnSe. Solid line is derived from the curve for the composite data set shown in figure 4a. The data (x) of Skinner[5] are essentially consistent with the predicted values given by the curve. The data (+) of Singh and Dayal[45] require a higher linear expansion coefficient than that predicted from the fit of the composite data set.

fits for four of these materials, GaAs, ZnTe, CdTe, and ZnSe, are illustrated in figures 1-4, respectively. Excellent agreement is found between predicted high-temperature lattice-parameter values and those experimentally measured by various workers (16,17,30,34, 35,36) for GaAs and ZnTe. As figures 3 and 4 illustrate, the high-temperature parameter predictions of the selected data agree with the high-temperature measured lattice parameters and provide justification for our selections. Excellent predictions are obtained at absolute temperatures as high as at least three times the Debye temperaure. Unfortunately, experimental measurements do not generally extend higher than that for the materials evaluated. Fits of data below 15 K are more qualitative and in this range are especially sensitive to inaccuracies in the data base. Data for AlSb and InAs are relatively limited and would be expected to have the least absolute accuracy of the data reported.

CONCLUSIONS

A phenomenological method for fitting thermal expansion and expansivity measurements based upon Blackman's physically significant but less rigorous definition of the Gruneisen constant has successfully predicted higher temperature thermal expansion and expansivity for a variety of Grimm-Sommerfeld compounds. We have also shown that a limited amount of higher temperature expansivity information can be helpful in evaluating the accuracy of lower temperature results. The method, because of its phenomenological nature, may be especially useful for predicting higher temperature properties of glasses and other disordered materials on the basis of lower temperature measurements. More accurate higher temperature thermal-expansion data for such materials and for industrial ceramics are important for the future design of composite systems. Additional work is planned to apply the method to such materials, especially those of lower symmetry.

REFERENCES

1. R. R. Reeber, Thermal expansion of some Group IV elements and ZnS, Phys. Status Solidi A 32:321 (1975).
2. T. H. K. Barron, J. G. Collins and G. K. White, Thermal expansion of solids at low temperatures, Adv. Phys. 29:609 (1980).
3. M. Blackman, On the thermal expansion of solids, Proc. Phys. Soc., London, Sect. B 70:827 (1957).
4. H. Adenstedt, Studien zur thermischen Ausdehnung fester Stoffe in tiefer Temperatur (Cu, Ni, Fe, Zinkblende, LiF, Kalkspat, Aragonit, NH_4Cl), Ann. Phys. 5(26):69 (1936).
5. B. J. Skinner, Thermal expansion of ten minerals, Geol. Surv. Prof. Pap. (U. S.) 450-D:D109 (1962).

6. R. R. Reeber and D. McLachlan, Jr., "Low temperature thermal expansion of wurtzite phases of IIB-VIB compounds. III. Thermal expansion of ZnS, sphalerite, from 4.2 to 333K and bibliography of thermal expansion of IIB-VIB compounds," Aerospace Res. Lab. Rep. No. 68-0183, Contr. No. AF33(615)-2280, Proj. No. 7885, USAF, AD-684559, Dayton, Ohio (1968).
7. J. S. Browder and S. S. Ballard, Low temperature thermal expansion measurements on optical materials, Appl. Opt. 8:793 (1969).
8. T. F. Smith and G. K. White, The low-temperature thermal expansion and Gruneisen parameters of some tetrahedrally bonded solids, J. Phys. C 8:2031 (1975).
9. B. J. Skinner, P. B. Barton and G. Kullerud, Effect of FeS on the unit cell edge of sphalerite, a revision, Econ. Geol. 54:1040 (1959).
10. R. S. Boorman and J. K. Sutherland, Subsolidus phase relations in the ZnS-In_2S_3 system, J. Mater. Sci. 4:658 (1969).
11. N. A. Goryunova and N. N. Fedorova, Solid solution in the ZnSe-GaAs system, Sov. Phys. - Solid State 1:307 (1959). [Translated from Fiz. Tverd. Tela 1:344 (1959).]
12. S. S. Ballard, S. E. Brown and J. S. Browder, Measurements of the thermal expansion of six optical materials, from room temperature to 250°C, Appl. Opt. 17:1152 (1978).
13. S. S. Devlin, J. M. Jost and L. R. Shiozawa, "Research on new high temperature semiconducting materials," Clevite Corp. Contr. No. AF33(616)-3923, WADD Tr60-11 (1960).
14. S. I. Novikova and N. Kh. Abrikosov, Thermal expansion of AlSb, GaSb, ZnTe, and HgTe at low temperatures, Sov. Phys. - Solid State 5:1558 (1964). [Translated from Fiz. Tverd. Tela 5:2138 (1963).]
15. L. R. Shiozawa, J. M. Jost, G. P. Chotkevys, S. S. Devlin and J. L. Barrett, "Research on II-VI compound semiconductors," Clevite Corp. 2nd Quarterly Rep. Contract No. AF33(657)-7399, USAF (1962).
16. H. J. Holland and K. Beck, Thermal expansion of zinc telluride from 0° to 460°C, J. Appl. Phys. 39:3498 (1968).
17. H. P. Singh and B. Dayal, Lattice parameters and thermal expansion of zinc telluride and mercury selenide, Acta Crystallogr. A26:363 (1970).
18. J. G. Collins, G. K. White, J. A. Birch and T. F. Smith, Thermal expansion of ZnTe and HgTe and heat capacity of HgTe at low temperatures, J. Phys. C 13:1649 (1980).
19. R. R. Reeber and C. Milton, George Washington Univ., unpublished data (1979).
20. J. S. Browder and S. S. Ballard, Thermal expansion measurements on four optical materials from room temperature to 10 K, Appl. Opt. 11:841 (1972).
21. M. G. Williams, R. D. Tomlinson and M. J. Hampshire, X-ray determinations of the lattice parameters and thermal

expansion of cadmium telluride in the temperature range 20-420°C, Solid State Commun. 7:1831 (1969).

22. L. I. Kleshchinskii, A. A. Makeev and P. V. Sharavskii, X-ray diffraction study of mercury selenide, Fiz. Tekh. Poluprov. 2:1002 (1968).

23. P. W. Davis and T. S. Shilliday, Some optical properties of cadmium telluride, Phys. Rev. 118:1020 (1960).

24. H. F. Schaake, "Study of thermal and mechanical properties of selected solids from 4 degrees to the melting point," Final Rep. No. AFCRL-69-0538 NTIS (U.S. Natl. Tech. Inf. Service) No. AD699579, 44 p. (1969).

25. T. Alper and G. A. Saunders, The elastic constants of mercury telluride, J. Phys. Chem. Solids 28:1637 (1967).

26. W. D. Lawson, S. Nielsen, E. H. Putley and A. S. Young, Preparation and properties of HgTe and mixed crystals of HgTe-CdTe, J. Phys. Chem. Solids 9:325 (1959).

27. E. Parthe, "Crystal Chemistry of Tetrahedral Structures," Gordon & Breach, New York (1964).

28. R. R. Reeber, unpublished data (1974).

29. V. I. Ivanov-Omskii, B. T. Kolomiets, L. I. Kleshchinskii and K. P. Smekalova, X-ray diffraction study of mercury telluride, Fiz. Tverd. Tela 10:3106 (1968).

30. N. Shaw and Y.-H. Liu, X-ray measurement of the thermal expansion of germanium, silicon, indium antimonide and gallium arsenide, Sci. Sin. 14:1582 (1965).

31. P. W. Sparks and C. A. Swenson, Thermal expansion from 2 to 40°K of Ge, Si, and four III-V compounds, Phys. Rev. 163:779 (1967).

32. S. I. Novikova, Investigation of thermal expansion of GaAs and ZnSe, Sov. Phys. - Solid State 3:129 (1961). [Translated from Fiz. Tverd. Tela 3:178 (1961).]

33. M. E. Straumanis and C. D. Kim, Phase extent of gallium arsenide determined by the lattice constant and density method, Acta Crystallogr. 19:256 (1965).

34. E. D. Pierron, D. L. Parker and J. B. McNeeley, Coefficient of expansion of gallium arsenide from -62 to 200°C, Acta Crystallogr. 21:290 (1966).

35. N. N. Sirota and Yu. I. Pashintsev, Determination of characteristic temperature and coefficients of linear expansion of arsenides of In and Ga, Inzhener. Fiz. Zhur., Akad. Nauk Belorus. SSR 1:38 (1958).

36. Yu. I. Pashintsev and N. N. Sirota, Temperature dependences of the characteristic temperatures and coefficients of linear expansion of arsenides of Al, Ga, and In, Dokl. Akad. Nauk Belorus. SSR, 3(2):38 (1959).

37. G. V. Ozolin'sh, G. K. Averkieva, N. A. Goryunova and A. F. Ievin'sh, An x-ray diffraction investigation of gallium and indium antimonides, Sov. Phys. - Crystallogr. 8:207 (1963).

38. G. Giesecke and H. Pfister, Prazisionsbestimmung der Gitter-

konstanten von AIIIBV Verbindungen, Acta Crystallogr. 11:369 (1958).

39. S. I. Novikova, Study of the thermal expansion of alpha-Sn, InSb, and CdTe, Sov. Phys. - Solid State 2:2087 (1961). [Translated from Fiz. Tverd. Tela 2:2341 (1960).]

40. H. E. Swanson, N. T. Gilfrich and G. M. Ugrinic, Standard x-ray diffraction powder patterns, Natl. Bur. Stand. (U. S.), Circ. 539 4:75 (1955).

41. N. N. Sirota and E. M. Gololobov, Atom scattering factors and electron density distribution in aluminum antimonide at 20 and 100°C, Dokl. Phys. Chem. 144:405 (1962).

42. H. E. Swanson, N. T. Gilfrich and G. M. Ugrinic, Standard x-ray diffraction powder patterns, Natl. Bur. Stand. (U. S.), Circ. 539, 4:72 (1955).

43. N. N. Sirota and L. I. Berger, Coefficients of linear expansion of arsenides of In and Ga and the tellurides of In and their relation to thermal conductivity, Inzhener. Fiz. Zhur., Akad. Nauk Belorus. SSR 2(5):104 (1959).

44. G. V. Ozolin'sh, G. K. Averkieva, A. F. Ievin'sh and N. A. Goryanova, An x-ray study of some A^3B^5 compounds which display deviations from stoichiometry, Sov. Phys. - Crystallogr. 7:691 (1963).

45. H. P. Singh and B. Dayal, X-ray determination of the thermal expansion of zinc selenide, Phys. Status Solidi 23:K93 (1967).

LATTICE THERMAL EXPANSION OF PRASEODYMIUM AND NEODYMIUM

J.V.S.S. Narayana Murty, R. Ramji Rao, and A. Ramanand

Physics Department
Indian Institute of Technology
Madras, India 600 036

ABSTRACT

The lattice thermal expansion of an ideal double hexagonal close-packed (dhcp) lattice is calculated using a nearest neighbor central force model and applied to the rare earth metals Pr and Nd, in which nearest neighbor central forces are predominant. The normal-mode Grüneisen parameters (GPs) are small, characteristic of the rare earth metals. It is found that, unlike the case of the ideal hcp lattice with nearest neighbor central interactions, there is anisotropy in the lattice Grüneisen functions $\gamma_{\perp}(T)$ and $\gamma_{\parallel}(T)$.

INTRODUCTION

The lighter rare earth elements praseodymium (Pr) and neodymium (Nd) crystallize in the double hexagonal close-packed (dhcp) structure with c/a values equal to 3.223 and 3.226 respectively, which are very near to the ideal value of 3.266. The theoretical expressions for the second-order elastic (SOE) and third-order elastic (TOE) constants of the ideal dhcp structure have been derived by Ramji Rao and Narayana Murty (1) using central interactions extending up to fourth neighbors. The Cauchy relation $C_{13} = C_{44}$ is nearly valid experimentally at all temperatures, as can be seen from the data of Griener et al. (2,3) for Pr and Nd. The ratio of the linear compressibilities $\chi_{\parallel}/\chi_{\perp}$ calculated from room temperature SOE constants gives the values 1.014 and 1.045 for Pr and Nd respectively. The ratio $\chi_{\parallel}/\chi_{\perp}$ is exactly one for the ideal dhcp lattice with nearest neighbor central interactions. These facts indicate that the forces in these metals are mostly central in character and farther neighbor interactions may be neglected. Hence in the present work the lattice thermal expansion coefficients of Pr and Nd have been calculated using a nearest neighbor central force model.

MODEL FOR THE LATTICE DYNAMICS OF THE STRAINED IDEAL dhcp LATTICE

The dhcp structure has an ABAC... stacking sequence of close-packed planes. The unit cell contains four atoms. This structure can be viewed as three interpenetrating sublattices A, B, and C. The environment of the atoms of the A sublattice is pseudo-cubic, while that of the atoms in B and C is hexagonal.

The potential energy is expanded in terms of the change in the square of the interatomic distance ΔR^2, i.e.,

$$\Phi = \frac{1}{2} \sum_{L,M,\lambda,\mu} \left\{ \frac{1}{2}\alpha(\Delta R^2)^2 + \frac{1}{6}\xi(\Delta R^2)^3 + \cdots \right\}$$

where α and ξ are the second and third derivatives of the potential energy respectively. Here λ and μ take values from 1 to 4.

The parameter α is obtained from the average of the values of C_{11} and C_{33}. The value of $(a^4/v)\alpha$ for Pr is 0.260×10^{11} dyn/cm^2 and for Nd it is 0.282×10^{11} dyn/cm^2. Here a is the lattice parameter in the basal plane, and v is the unit cell volume. The theoretical SOE constants along with experimental values (2,3) are presented in Table 1. The anharmonic parameter ξ is determined from the experimental compression data (4) using Murnaghan's equation of state (5) as described in reference 1. The pressure derivatives of the SOE constants are calculated using the expressions given by Ramji Rao and Srinivasan (6). The theoretical TOE constants and the pressure derivatives of the SOE constants are presented in Table 2 for both Pr and Nd.

The lattice is homogeneously deformed by the following strain components:

$$\eta_{xx} = \eta_{yy} = \tfrac{1}{2}\varepsilon', \qquad \eta_{zz} = \varepsilon'', \qquad \eta_{ij}(i \neq j) = 0$$

Here ε' is a uniform areal strain in the basal plane and ε'' is a uniform longitudinal strain along the unique axis. These strains do not create any internal strains (1) and hence do not violate the definitions of the thermal expansion coefficients. Using these strains the elements of the dynamical matrix are obtained and used to calculate the normal mode frequencies and the individual Grüneisen parameters (GPs).

EFFECTIVE LATTICE GRÜNEISEN FUNCTIONS OF Pr AND Nd

The temperature variation of the linear thermal expansion coefficients $\alpha_\perp$ and $\alpha_\parallel$ is expressed in terms of the effective Grüneisen functions $\gamma_\perp^\ell(T)$ and $\gamma_\parallel^\ell(T)$ as follows:

$$V\alpha_\perp = [(S_{11} + S_{12})\,\gamma_\perp^\ell(T) + S_{13}\,\gamma_\parallel^\ell(T)]\,C_v = \gamma_\perp^{Br}(T)\,C_v\,\chi_{iso}$$

$$V\alpha_\parallel = [2S_{13}\,\gamma_\perp^\ell(T) + S_{33}\,\gamma_\parallel^\ell(T)]\,C_v = \gamma_\parallel^{Br}(T)\,C_v\,\chi_{iso}$$

Here S_{ij} are the elastic compliance coefficients, V is the molar volume, C_v is the molar specific heat, and χ_{iso} is the isothermal

Table 1. SOE constants of Pr and Nd, in 10^{11} dynes/cm^2

C_{ij}	Pr		Nd	
	Theor.	Exp.	Theor.	Exp.
C_{11}	5.113	4.935	5.546	5.482
C_{12}	1.820	2.295	1.974	2.462
C_{13}	1.387	1.430	1.504	1.660
C_{33}	5.547	5.740	6.016	6.086
C_{44}	1.387	1.360	1.504	1.503

Table 2. Values of the anharmonic parameter ξ, TOE constants, and pressure derivatives of SOE constants of praseodymium and neodymium

Value in 10^{11} dyn/cm^2	Pr	Nd	$\frac{dC_{ij}}{dp}$	Pr	Nd
$(a^6/v)\xi$	- 2.045	-2.261			
C_{111}	-56.7	-62.6	$\frac{dC_{11}}{dp}$	7.00	7.17
C_{112}	-19.1	-21.1			
C_{113}	- 4.2	- 4.6	$\frac{dC_{12}}{dp}$	3.33	3.38
C_{123}	- 3.1	- 3.4			
C_{133}	-14.5	-16.1	$\frac{dC_{13}}{dp}$	2.45	2.50
C_{144}	- 2.7	- 3.0			
C_{155}	- 4.5	- 5.0	$\frac{dC_{33}}{dp}$	7.81	8.03
C_{222}	-68.5	-75.7			
C_{333}	-58.2	-64.3	$\frac{dC_{44}}{dp}$	1.44	1.50
C_{344}	-14.5	-16.1			

compressibility. $\gamma_{\perp}^{Br}(T)$ and $\gamma_{\parallel}^{Br}(T)$ are the average Grüneisen functions used by Brugger and Fritz (7).

The procedure of Blackman (8) is used to calculate $\gamma_{\perp}^{\ell}(T)$ and $\gamma_{\parallel}^{\ell}(T)$. The lattice frequencies are evaluated for the following sets of strains using the elements of the dynamical matrix:

(1)	$\varepsilon' = 0.0$	$\varepsilon'' = 0.0$
(2)	$\varepsilon' = 0.001$	$\varepsilon'' = 0.0$
(3)	$\varepsilon' = -0.001$	$\varepsilon'' = 0.0$
(4)	$\varepsilon' = 0.0$	$\varepsilon'' = 0.001$
(5)	$\varepsilon' = 0.0$	$\varepsilon'' = -0.001$

The corresponding frequencies are obtained as $\omega_1(q,j)$, $\omega_2(q,j)$, $\omega_3(q,j)$, $\omega_4(q,j)$, and $\omega_5(q,j)$ respectively, for a given wave vector q. The frequencies $\omega_1(q,j)$ are the normal-mode frequencies of the unstrained lattice. The GPs for any frequency are calculated as follows:

$$\gamma' = -\frac{1}{\omega_1(q,j)} \frac{\omega_2(q,j) - \omega_3(q,j)}{0.002}$$

$$\gamma'' = -\frac{1}{\omega_1(q,j)} \frac{\omega_4(q,j) - \omega_5(q,j)}{0.002}$$

The GPs $\gamma'(q,j)$ and $\gamma''(q,j)$ for the various normal-mode frequencies are evaluated over a grid of evenly spaced points (84) in the irreducible volume of the Brillouin zone. Dividing the frequency range into equal intervals of width $\Delta\omega = 0.08 \times 10^{13}$ rad/sec, the average of the GPs $\bar{\gamma}'(\omega)$ and $\bar{\gamma}''(\omega)$ is obtained for the frequencies in each interval. The $\bar{\gamma}'(\omega)$ and $\bar{\gamma}''(\omega)$ versus ω curves are shown for Pr and Nd in Figs. 1 and 2 respectively. In these figures, in the low temperature range $\bar{\gamma}'(\omega)$ and $\bar{\gamma}''(\omega)$ tend to their respective low-temperature limits calculated from the TOE constants using the procedure suggested by Ramji Rao and Srinivasan (9). The effective

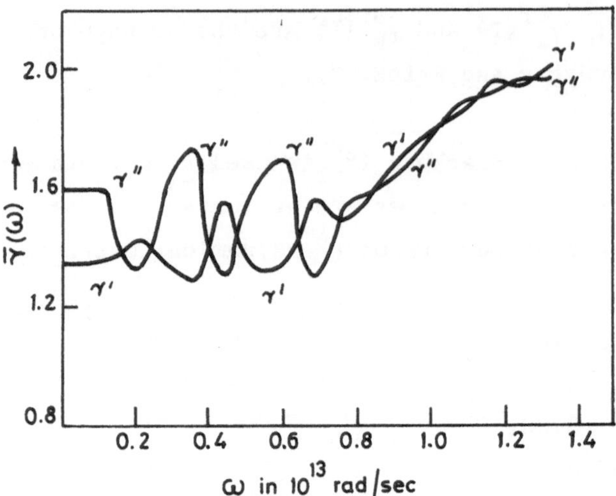

Fig. 1. Frequency variation of average GPs for praseodymium.

Grüneisen functions are then calculated as follows:

$$\gamma_{\perp}^{\ell}(T) = \int_{0}^{\omega_{max}} \bar{\gamma}'(\omega)\; g(\omega)\; \sigma(\omega,T)\; d\omega \Big/ \int_{0}^{\omega_{max}} g(\omega)\; \sigma(\omega,T)\; d\omega$$

$$\gamma_{\parallel}^{\ell}(T) = \int_{0}^{\omega_{max}} \bar{\gamma}''(\omega)\; g(\omega)\; \sigma(\omega,T)\; d\omega \Big/ \int_{0}^{\omega_{max}} g(\omega)\; \sigma(\omega,T)\; d\omega$$

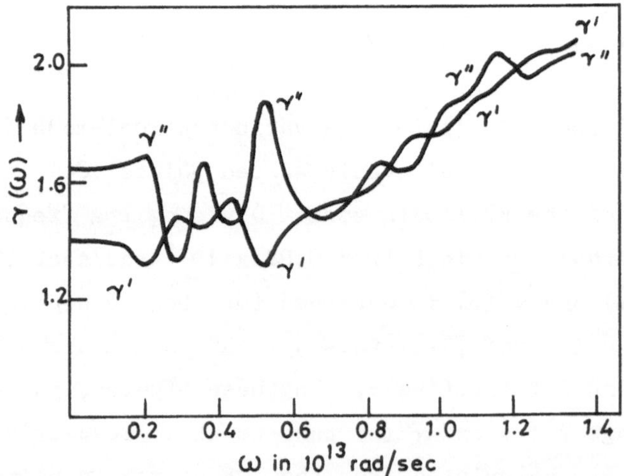

Fig. 2. Frequency variation of average GPs for neodymium.

Here $\sigma(\omega,T)$ is the Einstein's specific heat function and $g(\omega)$ is the frequency distribution function obtained by the root sampling technique. The volume Grüneisen function is obtained as

$$\gamma_V^{\ell}(T) = 2\gamma_{\perp}^{Br}(T) + \gamma_{\parallel}^{Br}(T)$$

The temperature dependences of $\gamma_{\perp}^{\ell}(T)$ and $\gamma_{\parallel}^{\ell}(T)$ are shown in Figs. 3 and 4 for Pr and Nd respectively. The temperature dependence of $\gamma_V^{\ell}(T)$ is similar to that of $\gamma_{\perp}^{\ell}(T)$ and hence is not presented. The frequency distribution functions $g(\omega)$ used in these calculations have been used to calculate the lattice specific heat of Pr and Nd (10).

DISCUSSION

The variation of the average GPs with ω for Pr and Nd is similar to that for the hcp rare earth metals (11). The high-temperature limit of the volume Grüneisen function $\gamma_V^{\ell}(T)$ has the value 1.69 for Pr and 1.75 for Nd. Ott (12) has measured the thermal expansion of Pr at very low temperatures. It is not possible to obtain the lattice contribution to γ from his measurements. Hence no comparison with experiment is possible.

It will be of interest to compare the ideal dhcp lattice with nearest neighbor central interactions with the ideal hcp lattice with nearest neighbor central interactions. The different stacking sequence, which causes the internal strain to contribute differently to the elastic constants, results in a different elastic behavior. For example, in the case of the hcp lattice (9), the relation $C_{144} = C_{155}$ is true though it is not a Cauchy relation, whereas it is not valid in the present case. At low temperatures, where acoustic modes dominate, this produces anisotropy in the lattice Grüneisen functions. However at higher temperatures the behavior is similar to that of the ideal hcp lattice (13), i.e., $\gamma_{\perp} = \gamma_{\parallel}$.

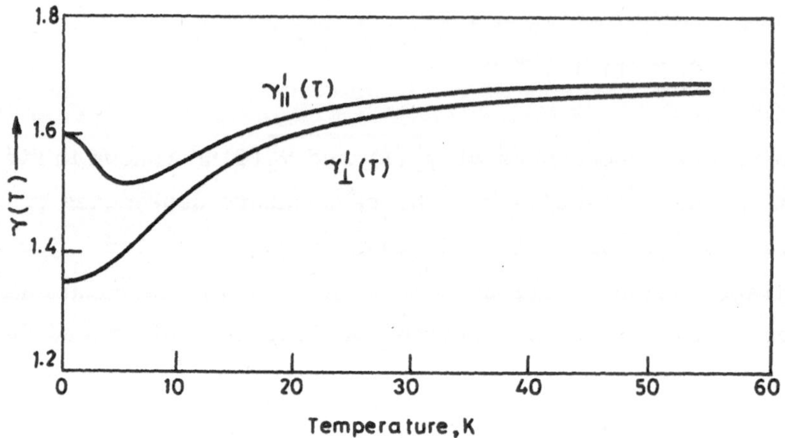

Fig. 3. $\gamma_{\perp}^{\ell}(T)$ and $\gamma_{\parallel}^{\ell}(T)$ versus T for Pr.

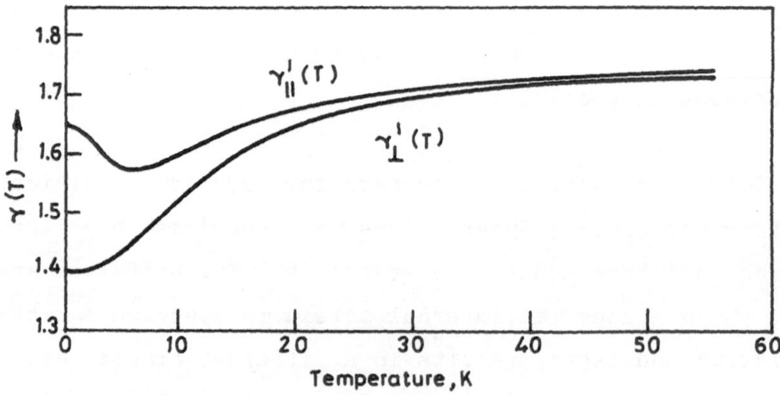

Fig. 4. $\gamma_{\perp}^{\ell}(T)$ and $\gamma_{\parallel}^{\ell}(T)$ versus T for Nd.

ACKNOWLEDGMENTS

The authors are grateful to the Council of Scientific and Industrial Research, Government of India, for the grant of research fellowships to Murty and Remanand.

REFERENCES

1. R. Ramji Rao and J.V.S.S. Narayana Murty, Physica 396, 194 (1979).
2. J.D. Griener, R.J. Schiltz Jr., J.J. Tonnies, F.H. Spedding, and J.F. Smith, J. Appl. Phys. 44, 3862 (1973).
3. J.D. Griener, D.M. Schlader, O.D. McMasters, K.A. Gschneidner, and J.F. Smith, J. Appl. Phys. 47, 3427 (1976).
4. "American Institute of Physics Handbook," Third Edition, Chapter IV, McGraw-Hill, New York (1972).
5. F.D. Murnaghan, Proc. Nat. Acad. Sciences, USA 30, 244 (1944).
6. R. Ramji Rao and R. Srinivasan, Phys. Stat. Solidi 31, K39 (1969).
7. K. Brugger and T.C. Fritz, Phys. Rev. 157, 524 (1967).
8. M. Blackman, Proc. Phys. Soc. (London) B70, 828 (1957).
9. R. Ramji Rao and R. Srinivasan, Phys. Stat. Solidi 29, 865 (1968).
10. J.V.S.S. Narayana Murthy and R. Ramji Rao, Indian J. Pure and Appl. Phys. 21, 623 (1983).
11. R. Ramji Rao and A. Ramanand, in "Thermal Expansion 6," edited by Ian D. Peggs, Plenum, New York (1978), p. 57; Acta Cryst. A33, 146 (1977).
12. H.R. Ott, Sol. Stat. Commun. 16, 1355 (1975).
13. R. Srinivasan and R. Ramji Rao, in "Inelastic Scattering of Neutrons," IAEA, Vienna (1965), Vol. 1, p. 325.

THERMAL EXPANSION ANISOTROPY OF CORUNDUM-TYPE STRUCTURE SINGLE CATION OXIDES

L. J. Eckert+ and R. C. Bradt

Ceramic Science and Engineering Program
Department of Materials Science and Engineering
The Pennsylvania State University
University Park, PA 16802

ABSTRACT

The thermal expansion of the seven common corundum structure oxides are examined with regard to the structural crystal chemistry concepts usually applied to thermal expansion trends in isostructural compounds. To explain the trends, it is necessary to classify these compounds into groups dependent on the nature of the individual bonding. The electronic interactions of transition metal cation pairs across the face shared octahedra strongly affect the c-axis thermal expansion coefficients. However, no similar structural effect is obvious to explain the trends of a-axis expansion coefficients.

INTRODUCTION

The thermal expansion of a solid is one of the most important physical properties. It is related to the nature of the individual bonding forces between the constituent atoms. If understanding is to be gained on a truly fundamental basis, it requires a rather strict, detailed theoretical approach, such as that advanced by Gruneisen(1) and Blackman(2). A less theoretical treatment is to examine the rather complex and only partially understood effects of structure on thermal expansion. This might be termed the empirical crystal chemistry approach. This avenue of understanding has been pursued extensively by Megaw(3), Bayer(4), Hummel(5), and

+Present address, Centre Engineering, 2820 E. College Ave., State College, Pennsylvania 16801

Rao[6] for different isostructural crystal types. It has achieved considerable success for several structure types including rutile, calcite, pseudobrookite, silica, dolomite, scheelite, and the alkali halides. These two approaches have been reviewed in detail by Collins and White[7] and by Kirchner[8], illustrating that both methods have achieved considerable success. In summary, it can be generalized that the theoretical methods are quite adequate and are most satisfactory if all of the appropriate physical constants are available for the calculations, but they rarely are. Similarly, the empirical crystal chemistry approach is quite satisfactory when the correct specific structural element is applied to the particular structure. Neither method is consistently capable of making a-priori predictions about different crystal structures, or even for compounds of a particular structure type. Additional studies of both approaches are clearly in order.

Publications of detailed x-ray thermal expansion data for several additional compounds of the corundum structure type[9-11] enable systematic analysis of the thermal expansion trends of that structure. The present paper collects and analyzes the axial thermal expansions of the seven common oxides of the corundum structure type (Al_2O_3, Cr_2O_3, Fe_2O_3, Ga_2O_3, Rh_2O_3, Ti_2O_3, and V_2O_3) and attempts to apply some of the previously successfully applied structural parameters to the members of that structure group. The results indicate that for the successful application of the crystal chemistry approach to thermal expansion, isostructural materials must often be further classified into sub-groups, dependent on the character of the individual bonding within the structure and specific structural features. The basis for this classification in the corundum structure is illustrated and the bonding interactions and structural features are noted.

Procedure

With the publication of detailed axial thermal expansion data for Rh_2O_3[9], Ga_2O_3[4,10], Ti_2O_3[11,17], and V_2O_3[11,18], there now exist excellent directional thermal expansion data for all seven of the common corundum structure oxides, including Fe_2O_3[12], Cr_2O_3[4,8,13], and of course Al_2O_3[14,15]. For the purpose of this study, the high pressure corundum types (In_2O_3, Tl_2O_3, etc.) reported by Shannon[16], and the "corundum-like" structures ($MgTiO_3$, $LiNbO_3$, etc.) referenced by Bayer[4] have not been considered, for much of the pertinent data on these materials is not available for an adequate comparison. Based on the historical use of the hexagonal indexing for alumina, the axial thermal expansions of these seven oxides were then expressed in terms of the a- and c-axes of the hexagonal unit cell, rather than the rhombohedral one. Figure 1 illustrates the orientation relation. Both the lattice parameters, and the thermal expansion coefficients of each of the

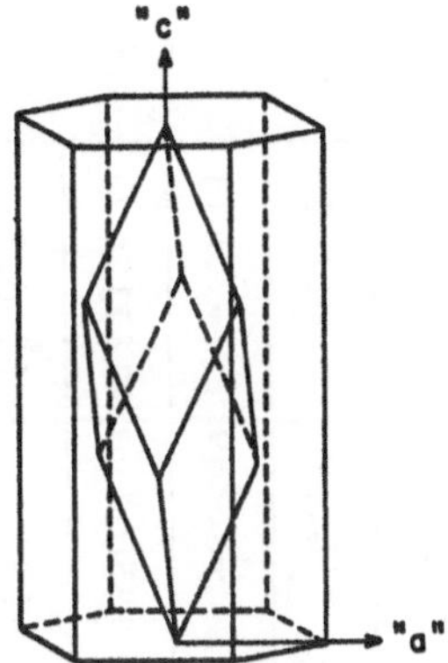

Figure 1. The orientation relation for the rhombohedral and hexagonal cells of corundum.

seven oxides were then transformed by computer regression analysis of the appropriate data into second order polynomials after Yaggee and Foote[19].

Results and Discussion

The seven single cation corundum structure compounds' axial coefficients of thermal expansion polynomials are listed in Table I. The lattice parameters are illustrated in Figures 2 and 3, with all data on the same scale to facilitate comparison. Evident in Figures 2 and 3, are the "relative" similarities of most of the compounds except Ti_2O_3 and V_2O_3, which possess negative thermal expansion coefficients, as well as inflections or minima in at least one of their lattice parameters. These two compounds are quite different in thermal expansion character from the other five compounds.

It is informative to compare the thermal expansion characteristics of these compounds on the basis of some of the structural concepts previously applied to analyze thermal expansion trends in isostructural compounds. Because of the similarity of these seven oxides, Megaw's valence and polyhedra tilting concepts cannot be readily applied[20]. These concepts appear to be most useful for comparing different structures and multi-cation compounds[21]. However, the cation sizes, bond lengths, lattice parameters, and their various related structural parameters would appear to be applicable.

TABLE I. THERMAL EXPANSION COEFFICIENTS OF THE CORUNDUM STRUCTURE OXIDES

	"α_a" = $X + YT + ZT^2$/°C			"α_c" = $X' + Y'T + Z'T^2$/°C		
	$X \times 10^6$	$Y \times 10^9$	$Z \times 10^{14}$	$X' \times 10^6$	$Y' \times 10^9$	$Z' \times 10^{14}$
Al_2O_3	+6.343	+3.166	-3.875	+7.116	+4.027	-5.685
Cr_2O_3	+8.209	-1.674	+1.771	+4.828	+3.147	-3.136
Fe_2O_3	+13.051	+0.805	-1.716	+7.545	+4.795	-7.358
Ga_2O_3	+5.232	+8.651	-10.42	+9.347	+6.591	-11.321
Rh_2O_3	+5.350	+1.281	-1.133	+5.246	+6.369	-7.480
Ti_2O_3(25-150)	+11.227	-237.0	-348.5	+5.503	+350.4	-1871
Ti_2O_3(200-300)	-91.16	+294.7	+761.9	+229.4	-689.0	+6006.8
Ti_2O_3(25-300)	-11.729	-70.02	-246.2	+34.677	+208.8	-2079.1
V_2O_3(25-400)	-	-	-	-12.727	+17.819	+24.05
V_2O_3(400-1000)	-	-	-	+1.131	+3.880	-2.220
V_2O_3(25-1000)	+32.664	-28.71	+77.47	-12.058	+22.582	+2.949

Directly related to cation size are the lattice parameters, which Bayer has shown to predict quite satisfactorily the thermal expansion coefficients for carbonates and nitrates with the calcite structure[22], for the dolomite-type borates[22], and for several of the pseudobrookites[23]. In each of these cases, the thermal expansion along one of the principal crystallographic directions varies linearly with that lattice parameter, although one must carefully choose the correct principal direction. This approach is not valid in the corundum structure. For example, ignoring Ti_2O_3 and V_2O_3 for the present, the largest parameters belong to Rh_2O_3 and the smallest to Al_2O_3. The α_a or α_c **values for** Cr_2O_3, Fe_2O_3, and Ga_2O_3 exhibit random intermediate values. Including both Ti_2O_3 and V_2O_3 with their negative expansions makes matters even worse. Although it is possible that over some narrow temperature range there might be some correlation, that temperature range is not obvious. Rather, one must conclude that the thermal expansion coefficients do not vary in a regular fashion with the lattice parameters in these corundum structure compounds.

Related to the lattice parameter approach is the bond length concept of Krishna Rao et al.[24,25], which has been successfully applied to explain the thermal expansions of TiO_2 and TeO_2. This concept is based on the premise that the strength of a bond can be related to its actual length in a particular crystal structure as compared to its ideal length in the same structure as determined from its ionic radii, coordination number, etc. Actual bond lengths that are shorter than the ideal are stronger and have lower thermal expansion coefficients, while correspondingly longer, weaker bonds should have higher thermal expansions. Kirchner[26] has applied similar bond strength type arguments to explain thermal expansion trends in the entire rutile structure family.

In the corundum structure, it is not clear which bonds are most important and how to evaluate them, primarily because of orientation effects and the accuracy of any "ideal" bond length calculations. However, numerous bonds have been identified [27,28] and measured [29,30] for all these compounds. Calculations of the ideal bond lengths, and comparison with the actual bond lengths, result in a complex situation overall; however, they do satisfactorily explain the Al_2O_3 and Ga_2O_3 thermal expansions. It fails miserably in several instances, namely in comparing the thermal expansion of Fe_2O_3, Cr_2O_3, Ti_2O_3, and V_2O_3, and also for comparing Rh_2O_3 with the others. The bond length concept appears to apply only to the oxides of Al_2O_3 and Ga_2O_3 within the corundum structure family and certainly not to the group as a whole.

The five compounds which do not ascribe to the bond length concepts, Cr_2O_3, Fe_2O_3, Rh_2O_3, Ti_2O_3, and V_2O_3, all contain transition metal cations, and the latter two of these have negative

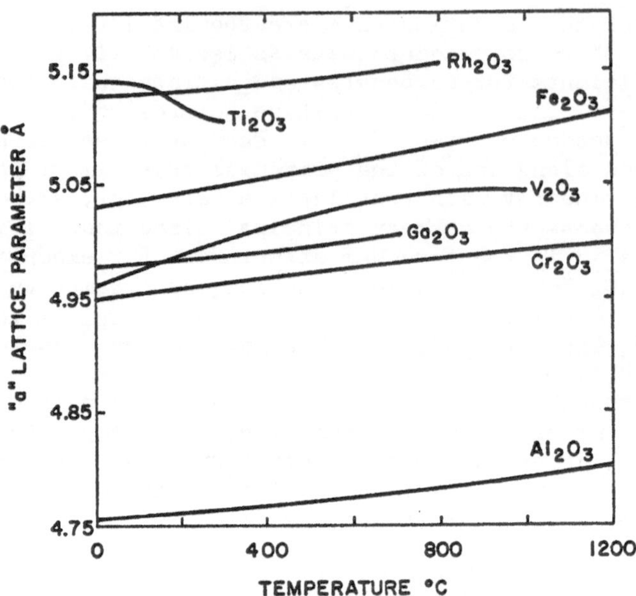

Figure 2. The "a" lattice parameters of the seven single cation corundum structure oxides as a function of temperature.

thermal expansion coefficients over some finite temperature range. Obviously, compared to Al_2O_3 and Ga_2O_3, there is something different about these five corundum structure oxides. Their thermal expansions suggest a grouping according to electronic structural interactions, such that Cr_2O_3 and Fe_2O_3 are similar, Ti_2O_3 and V_2O_3 are paired, and Rh_2O_3, the only member of the 4d transition series, is separate. The satisfactory bond length explanation of Al_2O_3 and Ga_2O_3 obviously suggests their similarity and pairing.

The pairing of Cr_2O_3 and Fe_2O_3 in thermal expansion trends is a natural coupling since both have a larger α_a than an α_c. At room temperature these two oxides have a thermal expansion anisotropy that is the reverse of that of Al_2O_3 and Ga_2O_3. Above 1200°C both Cr_2O_3 and Fe_2O_3 have reversed their thermal expansion anisotropy to agree with that of Al_2O_3 and Ga_2O_3. Above 1200°C the α_c's exceed the α_a's and both Cr_2O_3 and Fe_2O_3 also agree with the bond length principle. There is a reason for this anisotropy reversal and the lower room temperature expansion of Cr_2O_3 and Fe_2O_3 along the c-axis. Goodenough[31] points out that in the transition cation members of the corundum structure, the c-axis cation pairs exhibit a very strong d-electron interaction. This is because of the face sharing of octahedra along the c-axis. Figure 4 illustrates this unusual feature of the corundum structure. In the case of Al_2O_3

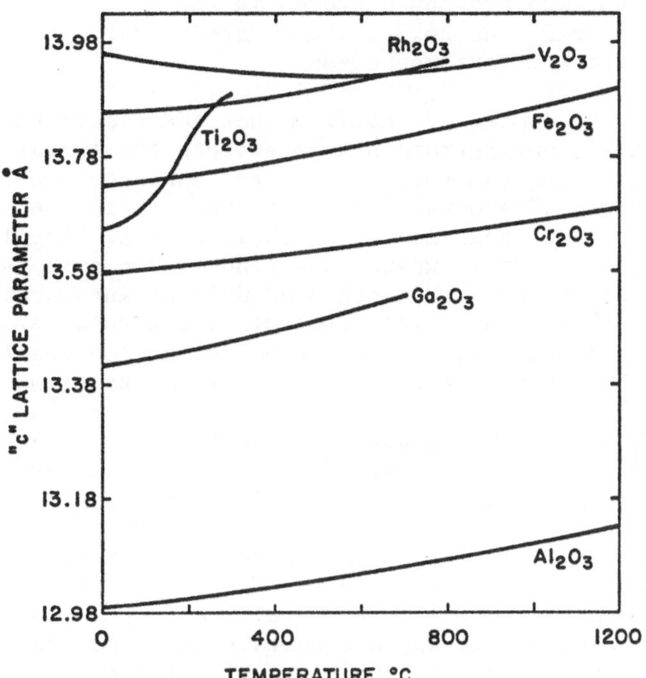

Figure 3. The "c" lattice parameters of the seven single cation corundum structure oxides as a function of temperature.

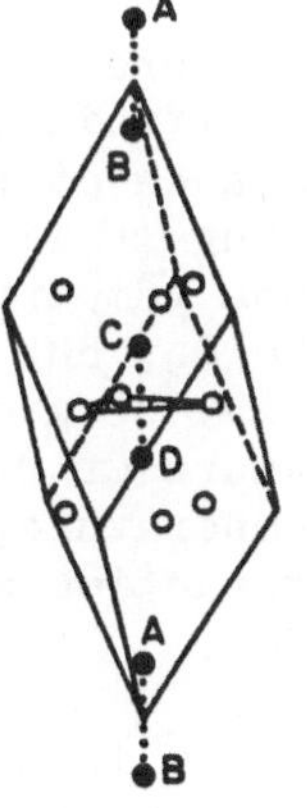

Figure 4. A schematic of some of the cation pairs which share common octahedral faces in the corundum structure. The triangle between the C-D pairs is one of the shared faces. Those with the A-B pairs and the anions involved are emphasized for clarity.

and Ga_2O_3, the cations in these face sharing octahedra actually repel each other and are not centered in the octahedra. However, in the case of Cr_2O_3 and Fe_2O_3, these cation pairs interact positively to create antiferromagnetism.

Greenwald and Smart[32] confirm that the electronic interaction leading to antiferromagnetism affects the thermal expansion. In the case of Cr_2O_3 and Fe_2O_3, it is the enhanced c-axis pairing across the shared octahedral faces that results in the reduced low temperature thermal expansion along their c-axes. At elevated temperatures, the normal expansion extends the c-axis cation pairs sufficiently to overcome this enhanced pairing interaction, so that above 1200°C, thermal agitation completely overcomes any antiferromagnetic coupling and the c-axes thermal expansion coefficients are then greater than those of the a-axes, similar to Al_2O_3 and Ga_2O_3.

Although neither Ti_2O_3 nor V_2O_3 are antiferromagnetic, Goodenough[31] reports that both also experience a very strong d-electron pairing along the c-axis. This results in a transition to metallic conduction because of delocalization of these d-electrons. In the case of Ti_2O_3, the electronic interaction and transition have been shown to be directly related to the observed thermal expansion inflection[11]. For V_2O_3 the correlation is less well defined; however, it appears certain that the negative c-axis expansion and the minimum at 500°C are also related to the c-axis electron interactions. However, the exact details remain to be fully explained.

A completely separate classification of the compound Rh_2O_3 could probably be justified on the basis of available data; however, its high temperature behavior strongly suggests that it can be grouped with Al_2O_3 and Ga_2O_3, as well as with the high temperature expansion of Cr_2O_3 and Fe_2O_3. At room temperature its bonds are longer than ideal and its thermal expansion coefficients are below those of Al_2O_3 and Ga_2O_3. However, the presence of the transition metal cation suggests caution in grouping it with Al_2O_3 and Ga_2O_3. Although it is nearly isotropic at room temperature, its anisotropy is the same as that of Cr_2O_3 and Fe_2O_3, but by 700°C an anisotropy reversal has occurred such that it may be grouped with Al_2O_3 and Ga_2O_3 as well as with high temperature Cr_2O_3 and Fe_2O_3. It might be speculated that below room temperature, Rh_2O_3 experiences some sort of strong cation-cation interaction along the c-axis, perhaps an antiferromagnetic transition.

Goodenough[31], also discusses the cation-cation interactions, perpendicular to the c-axis. Octahedra share edges in that orientation so some form of interaction is likely. Cation-anion-cation interactions may also be present. These certainly bear some responsibility for the Ti_2O_3 and V_2O_3 thermal expansions along the a-axis, however, the overall effects are not so pronounced as those along the c-axis.

Summary and Conclusions

The axial thermal expansions of the seven common single cation corundum structure oxides were systematically examined with regard to the commonly applied crystal chemistry concepts. The expansions of Al_2O_3, Cr_2O_3, Fe_2O_3, Ga_2O_3, and Rh_2O_3 are all of the same general magnitude, perhaps suggesting a very general structural relationship. However, in the vicinity of room temperature, only Al_2O_3 and Ga_2O_3 agreed with the bond length concept advanced by Krishna Rao et al.(24) The others differ because of some form of transition metal cation-cation electronic interactions. In the cases of Cr_2O_3 and Fe_2O_3, it is a *c*-axis antiferromagnetic pairing across the shared octahedral face that causes an anisotropy reversal in the thermal expansion coefficients. However, at elevated temperatures where thermal agitation overcomes the pairing interaction, both Cr_2O_3 and Fe_2O_3 agree with the bond length concepts and can be classified along with Al_2O_3 and Ga_2O_3. In a similar fashion Rh_2O_3 also exhibits a thermal expansion anisotropy reversal and agrees in principle with the bond length concept; however, no definitive evidence of a cation-cation interaction has been reported. Indications are that it may exist in Rh_2O_3 below room temperature.

Although Cr_2O_3, Fe_2O_3, and Rh_2O_3 clearly indicate that transition metal cation compounds should be considered separately, in any crystal chemical interpretation of thermal expansion, the negative thermal expansions of Ti_2O_3 and V_2O_3 further confirm this. The strong *d*-electron pairing of these two compounds results in a transition to metallic conduction as well as peculiar trends in the thermal expansion coefficients. Certainly these characteristics demand that they be classified separately so far as thermal expansion is concerned. The general deviant thermal expansion trends exhibited by the transition metal cation compounds of the corundum structure type suggest that transition metal containing compounds should be examined closely in any crystal chemistry scheme of thermal expansion trends.

REFERENCES

1. E. Gruneisen, Handbuch der Phys. 10, 1 (1926).
2. M. Blackman, Phil. Mag. 3, 831 (1958).
3. H. D. Megaw, Z. Krist. 100, 58 (1939).
4. G. Bayer, Proc. Brit. Cer. Soc. 22, 39 (1973).
5. F. A. Hummel, J. Amer. Cer. Soc. 32, 320 (1949).
6. K. V. Krishna Rao, p. 219, Thermal Expansion - 1973, Amer. Inst. Phys. Conf. Proc. No. 17, edited by R. E. Taylor and G. L. Denman, Amer. Inst. Phys., (1974).
7. J. G. Collins and G. K. White, p. 450, Prog. in Low Temp. Phys. IV, edited by C. G. Gartor, North-Holland Pub. Co., (1964).

8. H. P. Kirchner, p. 1, Prog. in Sol. St. Chem. I, Pergamon Press; (1964).
9. L. J. Eckert and R. C. Bradt, Mat. Res. Bull., 8, 375 (1973).
10. L. J. Eckert and R. C. Bradt, J. Amer. Cer. Soc. 56, 229 (1973).
11. L. J. Eckert and R. C. Bradt, J. App. Phys. 49, 3470 (1973).
12. A. T. Gordon, G. Bitsianes, and T. L. Joseph, Trans. AIME 233, 1519 (1965).
13. L. J. Eckert, M. S. Thesis, The Pennsylvania State University (1972).
14. W. J. Campbell and C. Grain, U. S. Bureau of Mines Report Investigation #5757 (1961).
15. J. B. Wachtman, Jr., T. G. Scuderi, and G. W. Cleek, J. Amer. Cer. Soc. 45, 319 (1962).
16. R. D. Shannon, Sol. St. Comm. 4, 629 (1966).
17. M. E. Straumanis and T. Ejima, Acta Cryst. 15, 404 (1962).
18. B. D. McWhan and J. P. Remeika, Phys. Rev. B 2, 3734 (1970).
19. F. L. Yaggee and F. G. Foote, Argonne National Laboratory Report #7644 (1969).
20. H. D. Megaw, Mat. Res. Bull. 6, 1007 (1971).
21. G. Bayer, J. Less Common Metals 26, 255 (1972).
22. G. Bayer, Z. Krist. 133, 85 (1970).
23. G. Bayer, J. Less Comm. Metals 24, 129 (1971).
24. K. V. Krishna Rao, S. V. Nagender, and L. Iyengar, J. Amer. Cer. Soc. 53, 124 (1970).
25. K. V. Krishna Rao and L. Iyengar, J. Mat. Sc. 7, 295 (1972).
26. H. P. Kirchner, J. Amer. Cer. Soc. 52, 379 (1969).
27. R. L. Blake, R. E. Hessevick, L. Foltai, and L. W. Finger, Amer. Mineral. 51, 123 (1961).
28. R. E. Newnham and Y. M. deHaan, Z. Krist. 117, 235 (1962).
29. M. Marezio and J. P. Remeika, J. Chem. Phy. 46, 1862 (1967).
30. J. M. O. Cory, Acta Cryst. B26, 1976 (1970).
31. J. B. Goodenough, p. 243, Magnetism and the Chemical Bond, Wiley and Sons (1963).
32. S. Greenwald and J. S. Smart, Nature (London) 166, 523 (1950).

THERMAL EXPANSION ANOMALY DUE TO RESIDUAL STRESS RELAXATION IN A POLYCRYSTALLINE ALUMINUM OXIDE

Y. Tree and D. P. H. Hasselman

Department of Materials Engineering
Virginia Polytechnic Institute and State University
Blacksburg, Virginia 24061 USA

ABSTRACT

A residually stressed coarse-grained polycrystalline aluminum oxide was observed to exhibit an irreversible change in dimension during the measurement of thermal expansion.

This effect was shown to be due to grain boundary cavitation and cracking within the tensile residual stress field by viscous deformation of the glassy grain boundary phase.

INTRODUCTION

Changes in temperature of materials generally are accompanied by dimensional changes. This phenomenon is referred to as thermal expansion. In the absence of changes in the character of atomic bonding and structural modifications, thermal expansion occurs in a continuous and usually monotonic manner.

The thermal expansion of solids can display discontinuous or irreversible behavior due to such effects as crystallographic transformations, crystallization or structural rearrangements in such materials as glasses. Decomposition reactions, densification of porous solids, creep deformation, the healing or growth of microcracks also can contribute to unusual thermal expansion behavior [1,2].

Aluminum oxide in the α-modification does not undergo a crystallographic transformation. For this material, thermal expansion data up to the melting point show a smooth monotonic change in length without discontinuities in either slope or absolute value [3].

In a recent study however, an irreversible thermal expansion effect was noted for samples of a residually stressed coarse-grained polycrystalline α-aluminum oxide. It is the purpose of this paper to report and explain these observations.

EXPERIMENTAL

The polycrystalline aluminum oxide used for this study was obtained from a commercial source.* Its density was approximately 3.81 $gms.cc^{-1}$. Figure 1 shows a typical SEM-fractograph of an as-received specimen broken at room temperature. The average grain size was near 25 μm with an occasional grain as large as 40 μm. The residual pore-phase existed primarily at grain boundaries or triple points. The principal impurities consisted of approximately 0.5% Si, 0.2% Ca and 0.03% Fe. Figure 2 shows a micro-probe distribution map of the Ca. The Si and Fe occupied the same position as the Ca. This suggests that these impurities exist in the form of a calcium-iron-alumino-silicate glass. As will be shown, this glass phase existed primarily at the grain boundaries or triple points.

The specimens were in the form of solid circular cylinders 0.504 cm in diameter by approximately 3.75 cms long. Residual stresses were introduced by a tempering treatment which consisted of quenching the specimens from about 1450°C into a silicon oil** at room temperature. This tempering treatment results in a residual stress state at thermal equilibrium at room temperature due to thermoviscoelastic effects during transient cooling. The residual stresses are compressive in the surface regions of the specimens and tensile in the interior [4]. The magnitude of the compressive residual stress in the surface was determined by measuring the fracture stress in three-point bending with a span of 1.27 cms. A set of 10 tempered specimens exhibited an average value of tensile fracture stress of 958 MPa with a coefficient of variation of 16.3 percent. For the as-received specimens the corresponding fracture stress was 545 MPa ± 9.4%. The difference in the average strength value of 413 MPa represents the magnitude of the compressive residual stress in the average specimen. The high value of the coefficient of variation of the tempered specimen compared to the as-received ones, suggests that the residual stress state can vary significantly from specimen to specimen. Possibly, this could be the result of differences in the

*AL-300, Western Gold and Platinum Co., Belmont, CA., USA.
**Dow Corning, Type 200, 100 cSt.

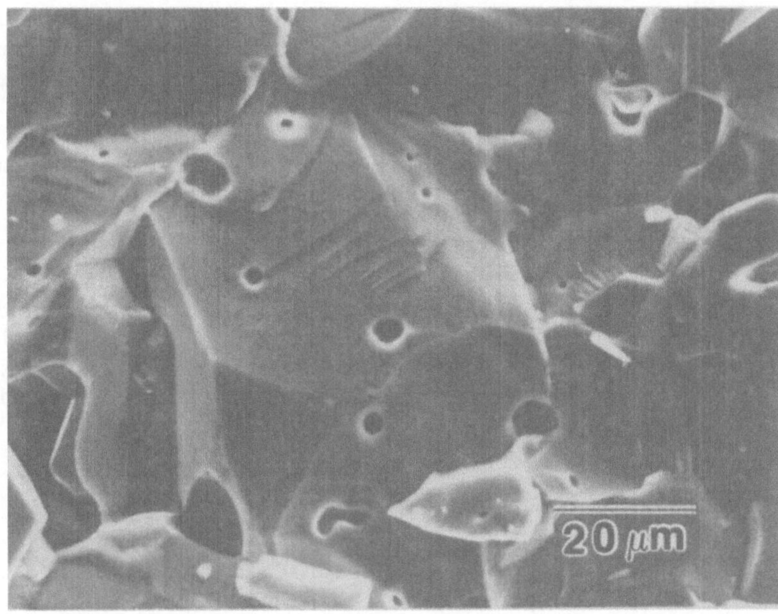

Fig. 1. Scanning electron fractograph of AL-300 polycrystalline aluminum oxide fractured at room temperature.

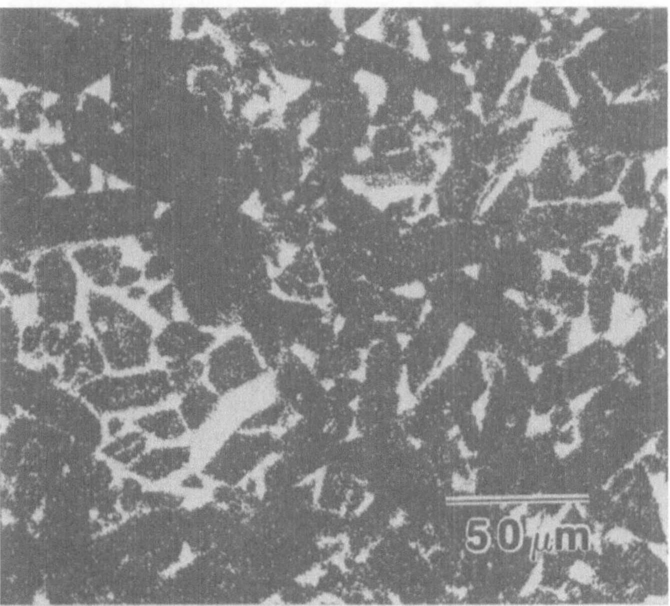

Fig. 2. Distribution of Ca impurities in AL-300 polycrystalline aluminum, coincident with Fe and Si impurities.

nature and amount of the glass phase at the grain boundaries. Because of the relatively high ratio of thickness-to-diameter, the strength values are somewhat overstated. Their relative values, however, indicate the existence of a residual stress field.

For the measurement of thermal expansion the as-received and tempered rods were cut to a length of approximately 2.2 cm. Thermal expansion data were obtained in a commercial dilatometer at heating and cooling rates of 10°C/min. A number of tempered specimens were held at a specified temperature within the dilatometer for a period of 25 min., preceded and followed by heating and cooling at rates of 10°C/min. These latter specimens were then fractured in three-point bending as discussed earlier for the purpose of determining the existence of residual stress relaxation and associated microstructural changes by SEM-fractography.

EXPERIMENTAL RESULTS AND DISCUSSION

Figure 3 show the thermal expansion data for an as-received and a typical tempered specimen of the aluminum oxide, respectively. The as-received sample shows the monotonic increase in specimen dimension with increasing temperature, in accordance with literature data [3]. It should be noted however that for this particular alumina the expansion data are below but within one percent of the recommended data for a high purity, dense and fine-grained polycrystalline alumina [3]. It is suggested that this effect is related to the anisotropy in thermal expansion of the individual grains. In this particular alumina, due to the presence of the glassy phase, the strains in the direction of maximum thermal expansion are accommodated by the formation of grain boundary cracks. The mechanism will be expected to be most effective for the range of temperature over which the glassy phase has sufficiently low viscosity for such grain boundary separation to occur. If this effect indeed occurred, the magnitude of the microstresses which result from the thermal expansion anisotropy in the alumina of the present study should be lower than those in a polycrystalline alumina without a glassy phase. In support of this conclusion it was found that the stress-sensitive R_1 and R_2 spectral lines emitted from Cr^{3+} impurities, for the present alumina showed significantly less broadening than for a dense aluminum oxide with no glassy phase. These latter observations are intended to be the subject of a future detailed report.

The thermal expansion for a specimen of the tempered alumina, as shown in Fig. 4, clearly shows a discontinuous behavior over the temperature range from about 850 to 950°C. Below and above this temperature range the thermal expansion for the as-received and tempered specimens are nearly identical. The slightly higher values of $\Delta L/L$ for the tempered specimen are thought to result

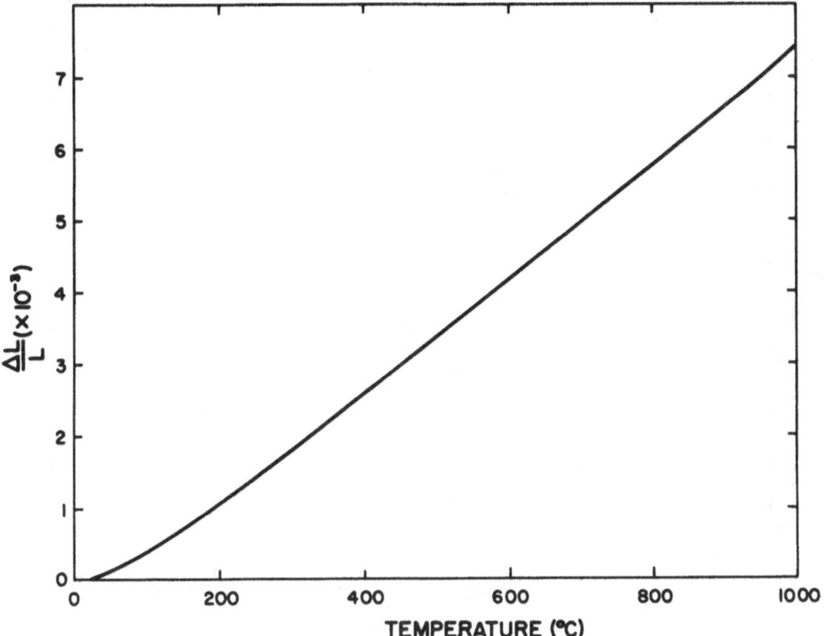

Fig. 3. Thermal expansion behavior of as-received AL-300 polycrystalline aluminum oxide.

from the reduced degree of grain boundary cracking due to the presence of the compressive stresses.

The discontinuous thermal expansion behavior for the tempered specimen from about 850 to 950°C is indicative of the existence of some phenomenon which operates in addition to the normal thermal expansion. Since this phenomenon is absent during cooling, the mechanism responsible appears to be irreversible and leads to a permanent increase in length on return to room temperature. Since the alumina was already in the α-form, this effect is not due to some irreversible crystallographic phase transformation.

It was observed further that tempered specimens, annealed for 25 min at 950°C showed a decrease in the fracture stress to an average value of 609 MPa for a total of three specimens. Such a decrease in fracture stress represents evidence for relaxation of the residual stresses introduced by the tempering process. No such decrease in fracture stress and stress relaxation was observed for 25 min anneals below 800°C. It appears then that the anomalous thermal expansion behavior shown in Fig. 4 is associated with the relaxation of the residual stresses. The irreversible

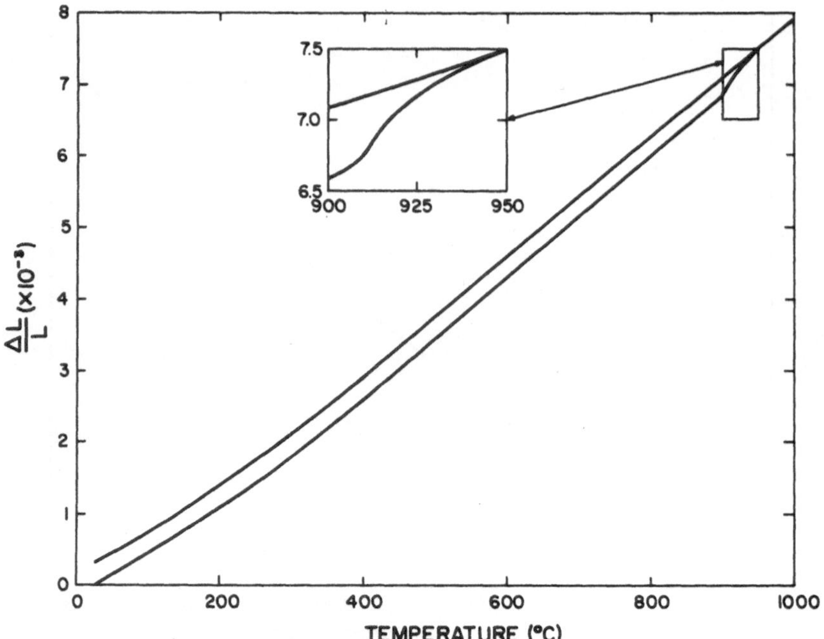

Fig. 4. Thermal expansion behavior of residually stressed AL-300 polycrystalline aluminue oxide.

increase in length of the tempered specimens indicates that this stress relaxation process is associated with the formation of cavities or cracks.

Stress relaxation processes by dislocation motion or diffusional processes such as Nabarro-Herring and Coble creep [5,6,7,8], exhibit approximately the same rates of creep in tension or compression. For this reason, apart from a possible effect of the multi-axial nature of the residual stress field, dimensional changes by these creep processes are expected to be minor, if not absent altogether. Furthermore, for an explanation of the present observations, the rate of creep deformation by these mechanisms is too slow by many orders of magnitude. Also, these creep processes, at least not directly, will not result in cavity or crack formation.

Creep by crack growth is thought to be the most likely mechanism responsible for the stress relaxation and irreversible increase in specimen dimension. Creep by crack growth is one of a number of possible mechanisms for elastic creep, which results from time-dependent changes in elastic properties [3]. Cracks in solids are very effective in lowering elastic moduli, which in turn provides the driving force for crack growth. Such crack growth can occur in high-purity polycrystalline aluminum oxide over the range of temperature of interest [9]. The presence of a glassy grain boundary phase, will accelerate such crack growth, as indicated by the recent theoretical results of Evans [10]. Elastic creep, which generally results in relatively small creep strains is particularly revelant to the present study, as complete residual stress relaxation can occur for creep strains no greater than the elastic strain due to the residual stresses.

Elastic creep by crack growth should occur primarily under conditions of tensile stress. This suggests that the residual stress relaxation and dimensional changes of the present samples occurred primarily by crack growth within the specimen interior, subjected to a state of tensile residual stress. Such creep by crack growth should be absent in the surface regions of the specimens which are in a state of compressive stress.

Experimental evidence for the validity of this hypothesis is given in Figs. 5a and 5b, which show SEM fractographs near the surface and within the interior, respectively, of a specimen held for 25 min at 850°C. Figure 5a as judged by the existence of cleavage patterns, is typical of transgranular fracture in aluminum oxide at room temperature. In contrast, the fractograph of Fig. 5b is indicative of intergranular fracture at higher temperatures. The existence of a glassy grain boundary phase is clearly evident, possibly as the result of incomplete wetting between the glass and the aluminum oxide. The geometry of the glass suggests it underwent viscous deformation during integranular crack formation. Literature data for the viscosity of a calcium aluminum silicate glass [11] suggest that with the level of residual stress in these specimens and temperatures in excess of 850°C considerably viscous deformation over a time period of 25 min. can occur.

For the above reasons, then, it appears reasonable to conclude that the glassy phase played a key role in the formation of the intergranular cracks. During fracture of the tempered and annealed specimen at room temperature, crack propagation through the interior regions followed the cracks generated at the higher temperatures leaving the features of the crack surface unchanged. Near the surface regions of the tempered specimens where no intergranular fracture occurred at the higher temperature, crack propagation occurred in the transgranular mode expected at room temperature.

Figure 6 schematically shows the originally internal stress distribution and the cracks in the specimen interior following the anneal at temperatures at which creep by crack growth is effective. No creep by any mechanism will occur in the regions of the specimens under compressive residual stresses. Nevertheless,

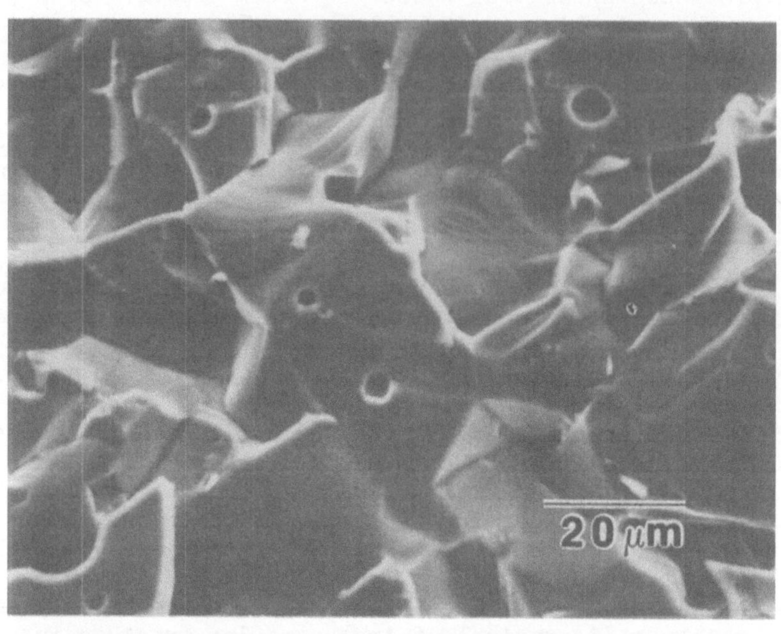

a

b

Fig. 5. Scanning electron micrographs of AL-300 tempered alumina subjected to anneal at 850°C for 25 min. a: Near specimen surface; b: Near specimen center.

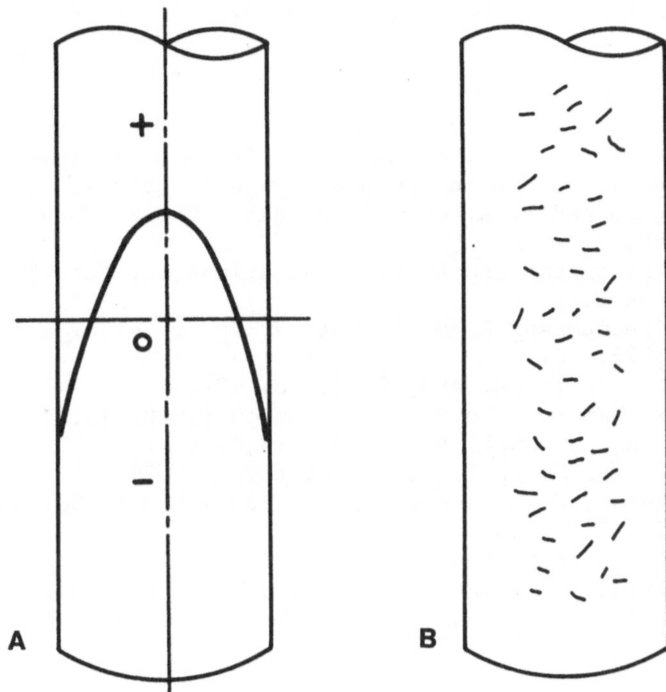

Fig. 6. Schematic presentation of A: original stress distribution and B: internal crack formation following anneal of residually stressed aluminum oxide.

due to the coupling between the tensile and compressive residual stresses, relaxation of the latter will occur entirely by elastic relaxation. The irreversible change in specimen dimensions shown in Fig. 4, should be equal to the release of strain for the elastic relaxation. An exact calculation of this effect is not possible as the residual stresses near the end of the specimen differ from those near the center

In general, the results presented in this study indicate that the relaxation of residual stress can make a contribution to the "apparent" value of the coefficient of thermal expansion. Such effects should be taken into account in the interpretation of the experimental data.

ACKNOWLEDGMENT

This study was supported by the Army Research Office, Research Triangle Park, N.C., under contract DAAG 29-79-C-0193.

A. Venkateswaran is acknowledged for many helpful discussions.

REFERENCES

1. W. D. Kingery, H. K. Bowen, D. R. Uhlmann, Introduction to Ceramics, 2nd Ed. John Wiley, N.Y. (1976).
2. E. A. Bush and F. A. Hummel, J. Amer. Ceram. Soc., 41 (1958) 189.
3. A. Venkateswaran and D. P. H. Hasselman, J. Mat. Sc., (in press).
4. W. R. Buessum and R. M. Gruver, J. Amer. Ceram. Soc., 55 (1972) 101.
5. J. Weertman, Trans. ASM, 61 (1968) 681.
6. F. R. N. Nabarro, Phys. Soc., London (1948) 75.
7. C. Herring, J. Appl. Phys., 21 (1958) 437.
8. R. L. Coble, J. Appl. Phys., 34 (1963) 1679.
9. A. G. Evans, M. Linzer and L. R. Russell, Mat. Sc. and Eng. 15 (1974) 253.
10. A. G. Evans, Acta Met., 28 (1980) 1155.
11. E. B. Shand, Glass Engineering Handbook, 2nd Ed., McGraw-Hill Book Company, Inc., N.Y. (1958) 20.

THERMAL EXPANSION OF FORSTERITE, Mg_2SiO_4

Isao Suzuki, Humihiko Takei,* and Orson L. Anderson

Institute of Geophysics and Planetary Physics
University of California at Los Angeles
Los Angeles, CA 90024, U.S.A.

*Research Institute of Iron, Steel and Other Metals
Tohoku University, Sendai, 980 Japan

INTRODUCTION

Forsterite Mg_2SiO_4 is the most important end member of natural olivine $(Mg,Fe)_2SiO_4$ which is a major constituent of the upper mantle of the Earth. Several investigators have grown forsterite single crystals for the measurement of physical properties of this mineral (Shankland, 1963; Takei and Kobayashi, 1974). Here, we report accurate measurements of the thermal expansion of synthetic single-crystal forsterite (See Table 1 and Figure 1). We compare these findings with the data of fayalite Fe_2SiO_4 (See Figure 2) and that of previous researchers (See Figure 3). Least squares fitting was made based on an equation derived by Grüneisen (1926) and the resulting values of these parameters in the equation are given in Table 2.

SPECIMEN

Sample crystals were grown by the Czochralski method (Takei and Kobayashi, 1974). This crystal is colorless and transparent, and does not contain any inclusion or cracks. Forsterite has orthorhombic symmetry and the crystallographic data of this crystal are as follows (See also, Syono et al., 1981): cell parameters a = 0.47516, b = 1.01920, and c = 0.59784 in nm, and molar volume V = 43.595 cm^3/mol at 25°C. The bulk density ρ_B = 3.225 g/cm^3 is close to X-ray density, ρ_X = 3.228 g/cm^3. The orientation of crystal axes were determined by the back Laue diffraction method

TABLE 1: Thermal Expansion of Forsterite

Temp. °C	Observed expansion Y_L (%)			Calculated expansion Y (%)				Calculated expansion coefficient $\alpha(10^{-6}/^{\circ}C)$			
	a	b	c	a	b	c	V	a	b	c	V
-200				-0.0705	-0.1223	-0.1321	-0.3215	0.3	0.6	0.9	1.74
-150				-0.0666	-0.1148	-0.1221	-0.3010	1.4	2.6	3.3	6.97
-100				-0.0561	-0.0957	-0.0995	-0.2499	2.8	5.1	5.7	13.45
-50				-0.0384	-0.0647	-0.0660	-0.1686	4.2	7.2	7.6	18.92
0				-0.0144	-0.0243	-0.0248	-0.0633	5.3	8.9	8.9	23.04
20				-0.0034	-0.0060	-0.0066	-0.0159	5.7	9.4	9.3	24.37
25	0.0000	0.0000	0.0000	-0.0005	-0.0012	-0.0019	-0.0037	5.8	9.5	9.4	24.68
50	0.0153	0.0218	0.0209	0.0145	0.0234	0.0221	0.0599	6.2	10.1	9.8	26.09
75	0.0297	0.0484	0.0454	0.0304	0.0493	0.0471	0.1267	6.5	10.6	10.2	27.33
100	0.0467	0.0764	0.0718	0.0472	0.0765	0.0730	0.1966	6.9	11.1	10.5	28.41
125	0.0643	0.1052	0.0990	0.0646	0.1046	0.0997	0.2690	7.1	11.4	10.8	29.37
150	0.0818	0.1351	0.1266	0.0828	0.1337	0.1270	0.3437	7.4	11.8	11.1	30.22
175	0.1013	0.1649	0.1562	0.1015	0.1637	0.1550	0.4205	7.6	12.1	11.3	30.99
200	0.1207	0.1945	0.1846	0.1208	0.1944	0.1835	0.4993	7.8	12.4	11.5	31.70
225	0.1406	0.2250	0.2132	0.1406	0.2257	0.2126	0.5798	8.0	12.7	11.7	32.34
250	0.1612	0.2575	0.2436	0.1608	0.2578	0.2421	0.6619	8.2	12.9	11.9	32.94
275	0.1818	0.2899	0.2735	0.1814	0.2904	0.2721	0.7455	8.3	13.1	12.1	33.50
300	0.2037	0.3221	0.3039	0.2025	0.3235	0.3026	0.8306	8.5	13.3	12.2	34.02
325	0.2246	0.3569	0.3346	0.2239	0.3572	0.3334	0.9170	8.6	13.5	12.4	34.52
350	0.2463	0.3918	0.3652	0.2457	0.3914	0.3646	1.0048	8.8	13.7	12.5	35.00
375	0.2680	0.4266	0.3964	0.2678	0.4261	0.3962	1.0938	8.9	13.9	12.7	35.46
400	0.2897	0.4615	0.4277	0.2902	0.4613	0.4282	1.1840	9.0	14.1	12.8	35.90
425	0.3128	0.4966	0.4594	0.3129	0.4969	0.4605	1.2754	9.1	14.3	12.9	36.33
450	0.3357	0.5337	0.4920	0.3360	0.5329	0.4932	1.3679	9.2	14.4	13.1	36.75
475	0.3591	0.5698	0.5244	0.3593	0.5694	0.5263	1.4616	9.4	14.6	13.2	37.17

TABLE 1 (continued)

Temp. °C	Observed expansion Y_L (%)			Calculated expansion Y (%)				Calculated expansion coefficient $\alpha (10^{-6}/^{o}C)$			
	a	b	c	a	b	c	V	a	b	c	V
500	0.3833	0.6062	0.5582	0.3829	0.6062	0.5597	1.5565	9.5	14.7	13.4	37.58
525	0.4062	0.6442	0.5917	0.4069	0.6436	0.5934	1.6525	9.6	14.9	13.5	37.98
550	0.4303	0.6806	0.6258	0.4311	0.6813	0.6275	1.7495	9.7	15.1	13.6	38.39
575	0.4552	0.7189	0.6610	0.4556	0.7194	0.6619	1.8477	9.8	15.2	13.8	38.79
600	0.4801	0.7586	0.6962	0.4803	0.7580	0.6967	1.9471	9.9	15.4	13.9	39.19
625	0.5054	0.7969	0.7319	0.5054	0.7969	0.7319	2.0475	10.0	15.5	14.0	39.60
650	0.5313	0.8367	0.7678	0.5307	0.8363	0.7674	2.1491	10.1	15.7	14.2	40.01
675	0.5563	0.8761	0.8045	0.5563	0.8761	0.8032	2.2519	10.2	15.9	14.3	40.42
700	0.5820	0.9147	0.8412	0.5822	0.9163	0.8395	2.3558	10.4	16.0	14.4	40.84
725	0.6088	0.9565	0.8775	0.6084	0.9570	0.8761	2.4609	10.5	16.2	14.6	41.26
750	0.6349	0.9976	0.9137	0.6349	0.9981	0.9131	2.5672	10.6	16.4	14.7	41.69
775	0.6616	1.0407	0.9515	0.6617	1.0396	0.9505	2.6747	10.7	16.5	14.9	42.13
800	0.6893	1.0819	0.9893	0.6888	1.0816	0.9882	2.7834	10.8	16.7	15.0	42.57
825	0.7164	1.1247	1.0273	0.7162	1.1241	1.0264	2.8935	10.9	16.9	15.2	43.03
850	0.7439	1.1669	1.0647	0.7439	1.1670	1.0650	3.0048	11.1	17.1	15.4	43.49
875	0.7722	1.2112	1.1038	0.7720	1.2104	1.1041	3.1175	11.2	17.3	15.5	43.97
900	0.8007	1.2538	1.1427	0.8003	1.2543	1.1435	3.2315	11.3	17.4	15.7	44.46
925	0.8283	1.2984	1.1821	0.8290	1.2987	1.1835	3.3470	11.5	17.6	15.9	44.96
950				0.8581	1.3437	1.2239	3.4639	11.6	17.8	16.1	45.48
1000				0.9172	1.4352	1.3061	3.7022	11.9	18.3	16.4	46.57
1050				0.9779	1.5290	1.3904	3.9470	12.2	18.7	16.8	47.73
1100				1.0402	1.6253	1.4770	4.1985	12.5	19.2	17.3	48.97
1150				1.1042	1.7242	1.5659	4.4574	12.8	19.7	17.8	50.32
1200				1.1701	1.8260	1.6574	4.7243	13.2	20.3	18.3	51.78

within an accuracy of ± 0.5°. The specimens were cut into rectangular prisms with length of 5-8 mm in three crystallographic orientations with a cross sectional area of approximately 3x3 mm^2.

RESULT

Thermal expansion was measured by a dilatometer in which the displacement detector is a differential transformer with high

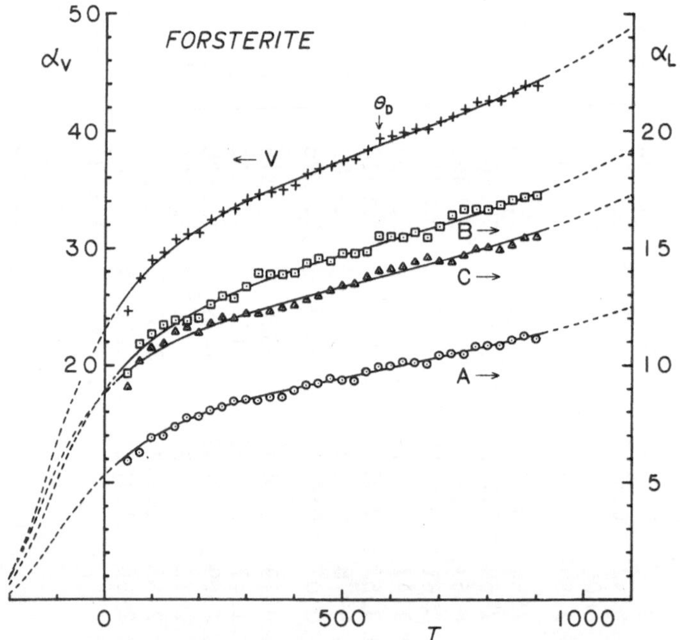

Fig. 1. Expansion coefficients α (in $10^{-6}/^oC$) vs. temperature (in °C) of forsterite, Mg_2SiO_4. A, B, C: a-, b-, c-axis, respectively, V: volume. Data points are obtained by numerical differentiation of expansion data Y_i with ΔT = 50°C. Smooth curves are calculated values by use of Equation (1) with parameters listed in Table 2. Θ_D is the Debye temperature determined from present data of Y_v.

sensitivity. Data acquisition was made with T ranging from room temperature to 925°C in temperature intervals of 25°C. Three relative expansions are calculated by $Y_i = \Delta L_i/L_i(25^oC)$, where L_i is the specimen length in each orientation; i.e., i = a-, b- and c-axis. Table 1 shows the average of Y_i data obtained by two or three cycles of heating and cooling runs. Reproducibility of the results is quite good and the difference between the cycles is less than one percent of the total elongation. The linear expansion coefficients are calculated by $\alpha_i = (1+Y_i)^{-1} (\Delta Y_i/\Delta T)$, where $\Delta T = 50^oC$. Volume expansion coefficients are obtained by

$$\alpha_v = \sum_i \alpha_i = \alpha_a + \alpha_b + \alpha_c$$

at each temperature. These data points show that the expansion coefficient in forsterite is a smoothly increasing function of temperature shown in Figure 1.

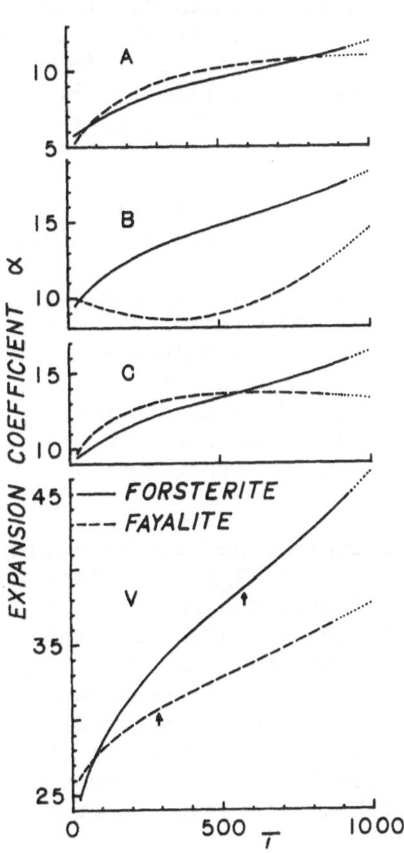

Fig. 2. Comparison of linear (a-, b-, and c-axis and volume (V) expansion coefficients α (in $10^{-6}/^oC$) of forsterite Mg_2SiO_4 and fayalite Fe_2SiO_4 in relation to temperature T (in °C). Arrows in the bottom figure indicate the Debye temperature.

DISCUSSION

Comparison of the present linear expansion Y_i data with that of previous research in the end member forsterite (Skinner, 1962; Smyth and Hazen, 1973; Hazen, 1976) show that these data have almost the same magnitude. The maximum difference is shown to be about 20 percent in each orientation. Present data of Y_b and Y_c are almost the same as that of Skinner (1962), but our results show Y_a as being systematically about 10 percent smaller than the results reported by Skinner. This systematic difference results in a smaller value in volume expansivity (about 3 percent).

Present Y_i data are very smooth functions of temperature in contrast to X-ray data, which is much rougher. This new curve is located in the middle of the other data (cf., same situation as volume expansion Y_v shown in Figure 3).

The linear expansion coefficient α_i is largest along the b-axis (direction of maximum linear compressibility) and the smallest along the a-axis (that of minimum linear compressibility), at least within the measured range of temperature. This situation is similar to that of tephroite Mn_2SiO_4 (Okajima et al., 1978) but different from that of fayalite Fe_2SiO_4 (Suzuki et al., 1981). A comparison of the expansion coefficient α_i(i = a, b, c and V) of forsterite and fayalite is shown in Figure 2, because they are the most important two end members of the major constituent of the Earth. The most significant difference between forsterite and fayalite is observed in the expansion coefficient along the b-axis. Usually, the temperature dependence of α_i is similar to those of the classical specific heat function, C_p or C_v. However, α_b in fayalite shows quite a peculiar temperature dependency, probably because of the antiferro-paramagnetic transition as well as some other factors (Suzuki et al., 1981).

It may be inferred from the composition-cell parameter or volume relationship that thermal expansion coefficients of olivine with composition $(Mg, Fe)_2SiO_4$ are located between those of both end members. The data of Rigby et al. (1941), show a systematic decrease of the expansion coefficient through the solid solution series from forsterite to fayalite. However, their values are small compared with the present data of forsterite and fayalite. We have examined many other published data on olivines since the work by Kozu et al. (1934), and found that data scatter so great that we cannot find any systematic trend from the experimental data, alone.

From the data derived from linear expansion, we have relative volume expansion, Y_v by $Y_v = (1+Y_a) \cdot (1+Y_b) \cdot (1+Y_c) - 1$. The present data of Y_v is compared with that of previous research in Figure 3. The X-ray data are still rough when compared with the dilatometric data. However, in the usual plot of Y_v vs. T, all data are so close to each other that they are shown by reducing a dominant factor, i.e., linear dependency on temperature, and $Z_v = Y_v - 0.003^* (T-25)$; Y_v in percentage and temperature (T) in ^oC (See Figure 3). Rigby et al., (1941) measured linear expansion of polycrystalline forsterite specimen with a reference temperature of 100^oC. Their data are shown with a common reference temperature of 25^oC, Y_V^R by use of an appropriate conversion based on an equation

$$Y_V^R = V(T)/V(25) -1 = (Y_v + 1) * (Y_v(100) + 1) -1$$

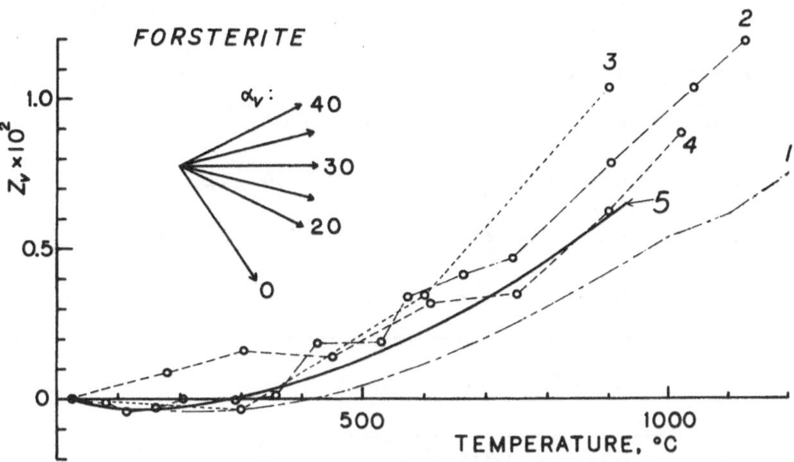

Fig. 3. Comparison of the present data with those of previous works. Volume expansion Y_V is transformed to Z_V by an equation: $Z_V = Y_V - 0.003*(T-25)$, (Y_V in %; T in °C) to show clearly the differences between them. Slopes in the upper left give the thermal expansion coefficients in 10^{-6}/°C. Data source: (1) Rigby et al. (1941); (2) Skinner (1962); (3) Smyth and Hazen (1971); (4) Hazen (1976); (5) present work. The Rigby, et al.'s data which is originally given by a reference temperature of 100°C is converted to those with the reference temperature of 25°C.

($Y_V = (Y_L + 1)^3 - 1$ is the volume expansion from Rigby et al.'s linear expansion data Y_L, and $Y_V(100)$ is the present data at 100°C, Y_V^R makes a smooth concave curve, but several percent smaller than the present ones in expansion coefficients. There are no data which coincide well with the present data in all measured ranges of temperature, i.e., one agrees well in the low temperature range and another in high temperature range, and present data lie almost at the average of the previous data. This situation also holds for Y_i.

Thermal expansion of many solid materials are well represented by the following equation (Suzuki et al., 1979).

$$k[y_V(T)]^2 - y_V(T) + E(\Theta,T)/Q = 0 \qquad (1)$$

where $y_V(T) = V(T)/V(o) - 1$ is the relative expansion of volume referred to 0 K, and $E(\Theta,T)$ is the thermal energy which may be evaluated by the Debye model using Debye temperature Θ and absolute

TABLE 2. Thermal Expansion Constants of Forsterite

	$Q/10^6 Jmol^{-1}$	Θ/K	k	$f/10^{-4}$	$s/10^{-4}$
a-axis	6.462 ± 0.026	951 ± 9	4.73 ± 0.06	7.11 ± 0.86	0.05
b-axis	4.192 ± 0.016	873 ± 9	3.06 ± 0.04	12.36 ± 0.15	0.08
c-axis	4.763 ± 0.032	738 ±17	3.41 ± 0.07	13.40 ± 0.30	0.12
Volume	4.975 ± 0.016	847 ± 8	3.61 ± 0.04	32.57 ± 0.34	0.17

$f \equiv L_i(25)/L_i(-273)-1$ or $V(25)/V(-273)-1$

$s \equiv \sqrt{\sum_{j=1}^{n} (Y_j^{obs} - Y_j^{calc})^2/(n-q)}$; n=37, q=4

temperature T, and two material constants Q and k. Equation (1) is derived for isotropic solids for which y_V is related to linear expansion y_L as

$$y_L = (y_V + 1)^{1/3} - 1. \quad (2)$$

Combining Equations (1) and (2), we can express linear thermal expansion y_L of isotropic materials by the same constants in Equation (1). This equation is also applied to a linear expansion of anisotropic solid (Suzuki et al., 1979). The least squares fitting of volume and linear expansion data in three crystallographic axes yielded the material constants and also additional constant f, which arises because the experimental reference temperature is 25°C, not 0 K. (See Table 2). The recalculated thermal expansion and its coefficients are shown in Table 1.

The Grüneisen's ratio $\gamma_{th} \equiv \alpha_V K_S V/C_p$ can be calculated by use of the value of the coefficient of thermal expansivity α_V together with the data of specific heat C_p (Stull and Prophet, 1971) and bulk modulus K_S (Sumino et al., 1977), and shown in relation to temperature (in Figure 4). Similar to many other minerals, the γ_{th} of forsterite is almost independent of temperature in the measured range of temperatures as shown in Figure 4. The γ_0 (supposed to be γ_{th} at 0 K) of forsterite (1.17), fayalite (1.13) and tephroite (1.08) are obtained from the definition of $Q \equiv K_0V_0/\gamma_0$ and also shown in Figure 4. It is observed that the Grüneisen's ratio of these olivines is almost the same in magnitude.

Comparing the thermal (γ_{th}) and acoustic (γ_{LT}, γ_{HT}) Grüneisen

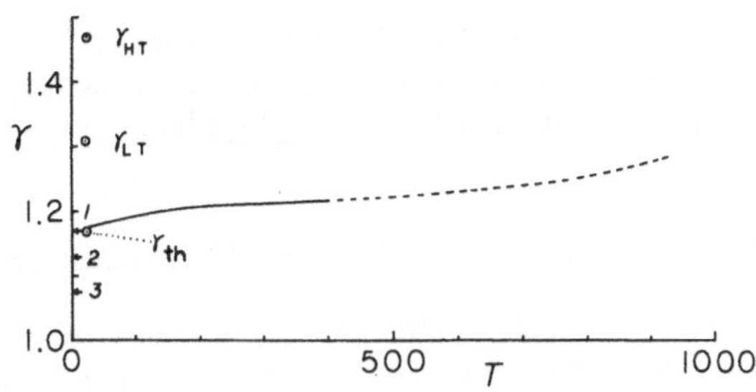

Fig. 4. Grüneisen's ratio vs. temperature, T, (°C), of forsterite. The smooth curve is thermal Gruneisen's ratio $\gamma_{TH} = \alpha_V K_S V/C_p$. Open circles show the acoustic (γ_{HT}, γ_{LT}) and thermal (γ_{th}) data by Kumazawa and Anderson (1969). Arrows indicate the γ_0 of olivines: (1) forsterite (present); (2) fayalite (Suzuki et al., 1981); (3) tephroite (Okajima et al., 1978) Dashed line above 400°C means extrapolation of K_S (Sumino et al., 1977).

parameters of forsterite, we can find some differences among them. The other constants Θ and k ($\equiv(\partial K/\partial P-1)/2$) in Table 2 are also just slightly different from those evaluated by the elasticity data: Θ = 760 K, k = 2.20 (Kumazawa and Anderson, 1969). The small but systematic differences between the values of γ and Θ obtained by thermal and acoustic data has important suggestion in geophysics (and will be discussed elsewhere.)

ACKNOWLEDGEMENTS

We acknowledge Professor M. Kumazawa of Nagoya University for his discussion and reading the manuscript. We also thank Dr. Y. Sumino for his help in the sample preparation. This work is supported by the National Science Foundation.

REFERENCES

Grüneisen, E. The state of a solid body, Handbuch der Physik, Bd. 10, Julius Springer (Berlin), pp. 1-52, 1926.

Hazen, R. M. Effects of temperature and pressure on the crystal structure of forsterite, Am. Mineral., 61, 1280-1293, 1976.

Kozu, S., Ueda, J., and Tsurumi, S. Thermal expansion of olivine, Proc. Imp. Acad. Japan, 10, 83-86, 1934.

Kumazawa, M. and Anderson, O. L. Elastic moduli, pressure derivatives, and temperature derivatives of single-crystal olivine and single-crystal forsterite, J. Geophys. Res., 74, 5961-5972, 1969.

Okajima, S., Suzuki, I., Seya, K., and Sumino, Y. Thermal expansion of single-crystal tephroite, Phys. Chem. Minerals, 3, 111-115, 1978.

Rigby, G.R., and Green, A. T. The thermal expansion characteristics of some calcareous and magnesian minerals, Trans. British Ceramic Soc., 41, 123-143, 1941.

Shankland, T. J. and Hemmenway, K. Synthesis of forsterite crystals, Am. Mineral., 48, 200, 1963.

Skinner, B. J. Thermal expansion of ten minerals, U. S. Geol. Surv. Prof. Paper, 450D, 109-112, 1962.

Smyth, J. R. and Hazen, R. M. The crystal structure of forsterite and hortonolite at several temperatures up to 900°C, Am. Mineral., 58, 588-593, 1973.

Stull, D. R. and Prophet, H. (ed.) JANAF-Thermochemical Table. (2nd ed.) Washington, D.C.: U. S. Dept. Commerce, National Bureau of Standards, 1971.

Sumino, Y., Nishizawa, O., Goto, T., Ohno, I., and Ozima, M. Temperature variation of elastic constants of single-crystal forsterite between -190° and 400°C, J. Phys. Earth, 25, 277-392, 1977.

Suzuki, I. Thermal expansion of periclase and olivine, and their anharmonic properties, J. Phys. Earth, 23, 145-159, 1975.

Suzuki, I., Okajima, S., and Seya, K. Thermal expansion of single-crystal manganosite, J. Phys. Earth, 27, 63-69, 1979.

Suzuki, I., Seya, K., Takei, H., and Sumino, Y. Thermal expansion of fayalite, Fe_2SiO_4, Phys. Chem. Minerals, 7, 60-63, 1981.

Syono, Y., Goto, T., Sato, J., and Takei, H. Shock compression measurements of single-crystal forsterite in the pressure range 15-93 GPa, J. Geophys. Res., 86, 6181-6186, 1981.

Takei, H., and Kobayashi, T. Growth and properties of Mg_2SiO_4 single crystal, J. Crystal Growth, 23, 121-124, 1974.

MEASUREMENT OF THE THERMAL EXPANSION OF FUSED SILICA

FROM -40°C TO 1000°C

Meihua Wang

The Institute of Metrology and Test of Chengdu
Chengdu, Sichuan, China

ABSTRACT

The measurement of the thermal expansion coefficient of fused silica from -40°C to 1000°C has been made using a laser interferometer and a unique multi-zone vacuum furnace.

For a specimen the average linear thermal expansion coefficient from -40°C to 1000°C is: α=47x10^{-8}/°C. This value is a mean value for twenty experimental data. Its root-mean square error is ±9x10^{-9}/°C.

METHOD

The thermal expansion coefficient α is derived from

$$\bar{\alpha}\,{}_{t_1}^{t_2} = \frac{\Delta L}{t_2 - t_1} \times \frac{1}{L} \tag{1}$$

where L is the length of specimen, t_1 and t_2 the temperature of the specimen, L_1 and L_2 the length of specimen at t_1 and t_2. When $t_2 \gg t_1$, the measurement of ΔL is very important for the sensitivity and the accuracy of the coefficient α. In our experiment, ΔL is measured directly by a He-Ne laser interferometer. The principle is shown in Fig. 1. The difference of optical path between the two surfaces is

$$r=2nL\cdot COS\theta + \frac{\lambda}{2} \tag{2}$$

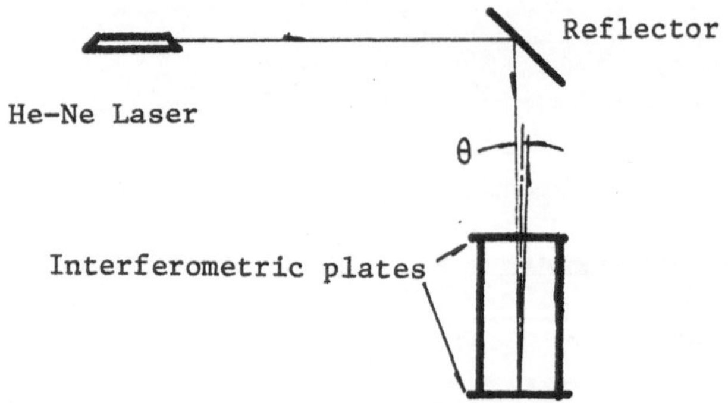

Fig. 1. The principle of the optical system.

where n is the index of refraction of the atmosphere between the surfaces. In vacuum, n=1. L is the distance, in our apparatus it is the length of specimen. θ is the angle between the direction of the incident rays and the normal to the surface. λ is the wave length of the light source. The difference of optical path between the surfaces can be expressed with N--the number of contained waves between the surfaces:

$$r = N \cdot \lambda \tag{3}$$

If eqs. (2) and (3) are integrated,

$$N \cdot \lambda = 2L \cdot \cos\theta \pm \frac{\lambda}{2} \tag{4}$$

then

$$\Delta L = \Delta N \cdot \frac{\lambda}{2} \cdot \frac{1}{L \cdot \cos\theta} \tag{5}$$

where N is the difference of the order of interface and also the number of fringes that pass a fiducial mark. Finally, the thermal expansion coefficient can be expressed with this formula:

$$\bar{\alpha}_{t_1}^{t_2} = \frac{\Delta N}{t_2 - t_1} \cdot \frac{\lambda}{L \cdot \cos\theta} \tag{6}$$

APPARATUS

The experimental apparatus consists of two parts - the optical system and the -40°C to 1000°C vacuum furnace.

The Optical System

The optical system is shown in Fig. 2. The He-Ne laser is used as a light source. The rays partially reflect from the beam splitter into the interferometer, which sits in the vacuum furnace. Then, through the beam splitter and a diverging lens, the interference pattern falls upon a photoelectric cell with an adjustable slit. The photovoltage from the photocell is introduced into an amplifier. At last, a cyclic voltage signal will be recorded by a recorder, as the length of the specimen changes with the furnace temperature. Every waveform made by the recorder corresponds to one interference order change, also one fringe.

At the same time, the interference pattern is reflected by the beam splitter and falls upon the screen. Thus one can observe the change of shape and position of the interference pattern.

In our experiment, the interferometer consists of two sapphire plates and a tube specimen. Their diameters are about 30 mm. The length of the specimen is about 50 mm.

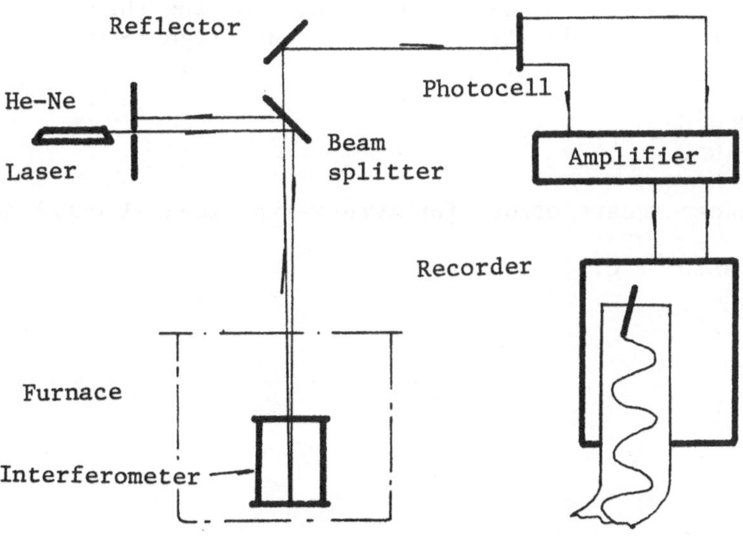

Fig. 2. The optical system.

In order to evacuate the air from the inside of the tube specimen, three channels were ground at equal intervals around the circumference of one end of the tube. The two sapphire plates used as the interferometer plates are uncoated.

<u>-40°C to 1000°C Vacuum Furnace</u>

There are three features on the vacuum furnace:

1) A stainless steel cooler is arranged between the specimen and the heater. When liquid nitrogen flows over the cooler, the specimen will be cooled.

2) The specimen is placed or taken off from the optical plates directly.

3) The heater consists of five windings of nickel-chrome wire with different lengths. The wires are wound on the inside wall of a refractory ceramic tube. The maximum power of the heaters is about 4000w. This arrangement of the heaters allows for both heating the specimen and distributing the power to each winding in order to decrease the axial temperature gradient.

RESULTS

The results in this paper were obtained under different conditions, i.e., the specimen was placed and taken off again and again, the optical system was also adjusted again and again, the rate of heating or cooling of the specimen was changed and the metering device was changed. The results of the thermal expansion coefficient of fused silica in the range from -40°C to 1000°C are

$$\bar{\alpha}\,{}^{1000°C}_{-40°C} = 47\times10^{-8}/°C$$

The root-mean square error (for twenty experimental data) is

$$\Delta\alpha = \pm 9\times10^{-9}/°C.$$

THERMAL EXPANSION BEHAVIOR OF PLASMA-SPRAYED OXIDE COATINGS

S. Rangaswamy*, S. Safai**, H. Herman***

*METCO Inc., 1101 Prospect Avenue, Westbury, Long Island, New York 11590; **Pratt & Whitney Aircraft Group, United Technologies, West Palm Beach, Fla. 33402; ***State University of New York, Dept. of Materials Science & Engineering, Stony Brook, New York 11794

ABSTRACT

Plasma-sprayed protective coatings are used in a variety of applications to prevent the degradation of metallic substrates due to high temperature oxidation, corrosion, and wear. While a wide range of metals and ceramics can be deposited on metallic substrates by plasma spraying, the integrity of ceramic coatings is significantly affected by their relatively low thermal expansion coefficients and brittle fracture behavior. The difference in thermal expansion coefficient between the ceramic coating and metallic substrate is even more important in high temperature thermal cycling applications, for example, in industrial/utility gas turbine hot-sections and in automotive/marine diesel engine components.

It is well known that bulk ceramics (sintered, hot pressed) require a low thermal expansion coefficient for thermal shock resistance. This is generally not the case for plasma sprayed ceramic coatings, where the matching of the expansion coefficients between coating and substrate is desirable, yielding improved coating adhesion.

This paper focuses on a study of thermal expansion behavior of plasma sprayed Al_2O_3/TiO_2 and stabilized ZrO_2 compositions, substrate free, and with various substrate adherence conditions. A modified, single push-rod dilatometer was used to measure the expansion coefficients of plasma-sprayed coatings. Effects due to variations in porosity, composition, and thermal cycling are discussed.

INTRODUCTION

Oxide coatings, formed by plasma spraying onto properly prepared metallic substrates, adhere tenaciously under extreme service environments. Of special importance, and of key significance in aerospace applications, for example, is the need for the coating to possess adequate thermal fatigue resistance, maintain low thermal conductivity, and protect the metal substrate from chemical and mechanical attack.

Under conditions of thermal cycling, where the expansion coefficients between coating and substrate differ considerably, large stresses develop within the coating; tensile for the heating and cooling rates, as well as the extremes of temperature, and the time at the high temperature, can be determining factors in the performance of a ceramic coating system.

For strongly bonded coatings, the higher expansion of metal substrates is compensated by the formation of lamellar and radial (transverse) microcracks within the oxide coating. With continued thermal fatigue, the coating may crack into segments and will "follow" the expanding substrate. Likewise, on cooling, cracks will actually close and, at least visually, it might appear that the system has healed the damage which was introduced during the primary heating cycle. Such, however, cannot actually be the case.

It is thought that in certain oxide systems, transformation-induced microcracking (e.g., partially-stabilized zirconia) will relieve residual stresses and permit the coating to expand together with the underlying substrate during heating. It may further be possible to design transformations which will give rise to microcracking to an extent such that large cracks are blocked and are not able to propagate. These concepts, together with the segmented coating design approach, will, it is envisioned, give rise to new and highly effective coatings.

At the center of the problem of designing a coating system for thermal cycling applications resides the need to know the differential expansion coefficients between coating and substrate. One obvious goal is to spread out the sharp strain discontinuity at the interface through the use of graded coatings. This involves either continuously changing the chemistry of the coating, from the interface out, or by layering the coating. The net effect is to distribute the thermal expansion mismatch, leading to stress relief and, thus, to improved reliability.

Clearly needed are measurements of the differential thermal expansion coefficients and techniques for minimizing this differential. Or, if stress reduction is not feasible there should be developed means for designing-in relief schemes for handling the stresses.

While there exists published data on the thermophysical characteristics of bulk ceramics, such data cannot be directly utilized in the design of plasma-sprayed systems. There is very limited literature available on the thermophysical properties of plasma-sprayed coatings. This is due mostly to the paucity of methods available for non-destructive evaluation of protective coatings, especially in a high temperature environment.

This paper focuses on the importance of thermal expansion of plasma-sprayed oxide coatings and of some experiments carried out to understand how thermal expansion determines coating survivability.

EXPERIMENTAL

The linear thermal expansion coefficients were determined for plasma-sprayed coatings of alumina with varying amounts of titania (wt. % 0, 2.5, 13, 40), pure titania, and partially stabilized zirconia (CaO or Y_2O_3). Expansion measurements were made using a single silica push rod dilatometer. The modified sample holder used with the dilatometer and experimental details are described elsewhere[1]. The coefficients of expansion were determined for substrate-free coatings of the above compositions as well as coatings bonded to the substrates. In the latter case, the expansivity of the coating only is measured, but the measurement will, of course, be influenced by the substrate. This measurement was carried out to determine how the adhesion to a higher expansivity substrate will influence the α of the oxide overlay. In addition, coefficients of expansion were determined for different substrate preparations, such as grit-blasted substrates and adhesion-enhancing intermediate bond-coats (nickel-aluminide). The effect of porosity on the coefficients of expansion was also determined.

To correlate the thermal expansion characteristics to the thermal shock resistance, the coatings as attached to the substrates were thermally cycled to 1000°C in an argon atmosphere and the resulting cracking/spalling characteristics were studied.

RESULTS AND DISCUSSION

Published data on the thermal expansion of oxides (fused, hot pressed, etc) generally show considerable scatter. However, some general conclusions can be drawn from literature:

The work done by Kingery (1957) on sintered ceramics showed

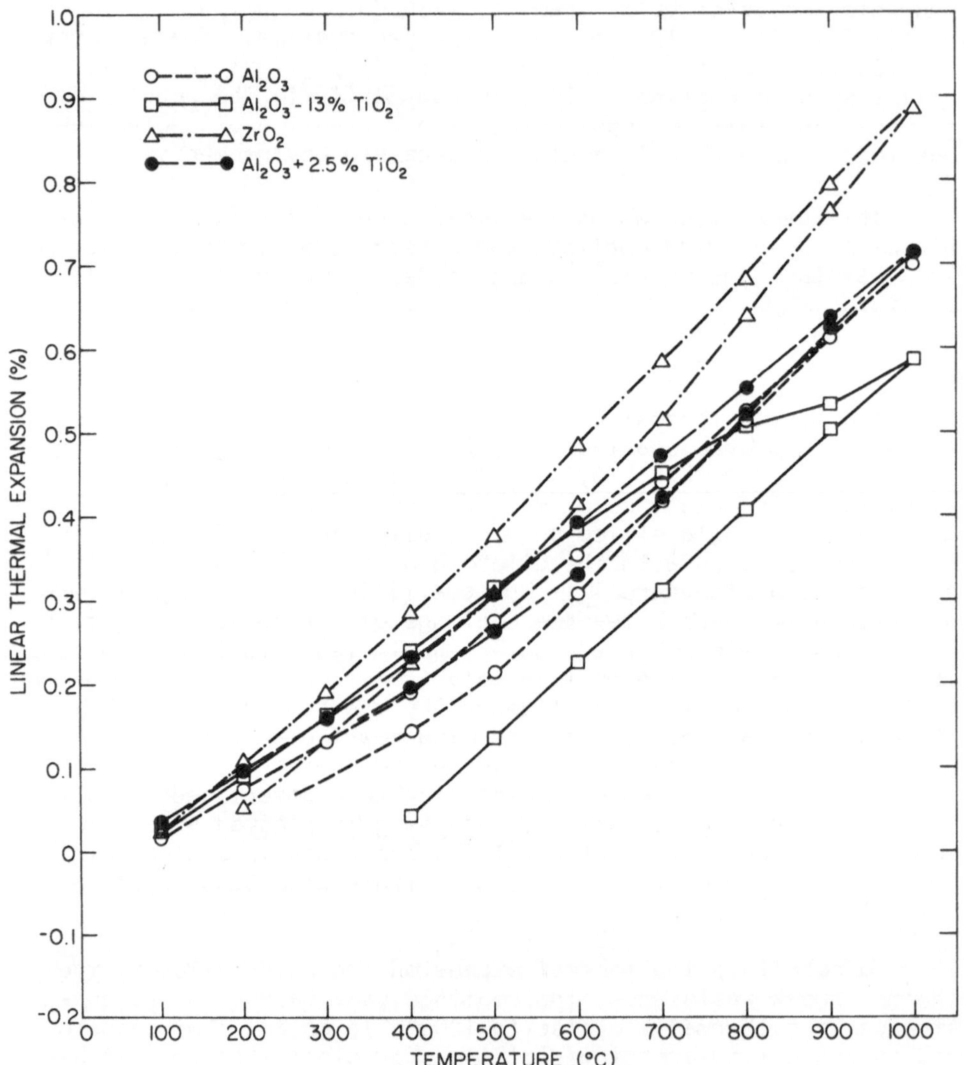

Figure 1: Linear thermal expansion percent of substrate free oxide coatings.

that the expansivity of two phase systems is not simply the average of the end-member values, but agrees quantitatively with calculations based on the assumption of substantial microstresses resulting from constraints of each phase on cooling. This process is known to give expansion hysteresis in systems such as Al_2TiO_5 and TiO_2. There is also evidence for extreme anisotropy in the thermal expansion of the alumina titanate phase, which can generate internal stresses so high that rupture can occur[2].

The results of expansivity of the substrate-free coatings in the Al_2O_3/TiO_2 system and ZrO_2 (stabilized) indicate that the expansivities of Al_2O_3, TiO_2, and ZrO_2, although slightly different from those values measured on pure fused oxides[3], have essentially the same relative ranking. Figure 1 shows the linear thermal expansion of the substrate-free oxides as a function of temperature up to 1000°C. The significant shrinkage generally observed during the cooling cycle is attributed to pore closure and microcrack healing in the oxide coating. Considerably larger shrinkage can occur with isothermal heat treatment at elevated temperatures. In this study, however, samples were only cycled to 1000°C without any isothermal holds. In the Al_2O_3 - TiO_3 system, however, the average coefficient of expansion shows a minimum in coatings containing up to 15 wt. % TiO_2. Some explanation regarding this phenomenon can be given from the previous microstructural investigation of plasma-sprayed Al_2O_3 - TiO_2 coatings (Ref. 4). TEM and x-ray diffraction examinations have revealed the formation of Al_2TiO_5 in Al_2O_3 - TiO_2 coatings, and, in particular, in 13-15 wt % TiO_2 compositions. The Al_2TiO_5 is known to exhibit significant anisotropic expansion (hysteresis) and also, because of low ultimate strength, to undergo microcracking. This effect is clearly evident in the cyclic linear thermal expansion curves for Al_2O_3 - 13 wt % TiO_2 shown in Figure 1.

The minimum in expansivity can be attributed primarily to the formation of microcracks in Al_2TiO_5 phase along the Al_2O_5 particle boundaries within the oxide coatings. The findings of Kingery that the expansivity of a two-phase system can show variations from the average of the end member values also support this observation.

Effect of Porosity

In an earlier investigation by Coble and Kingery[5] on the influence of porosity on thermal expansion of sintered bulk Al_2O_3, it was found that the thermal expansion coefficient does not appreciably vary with porosity. Similarly, Austin[6] reported that porosity affects thermal expansion negligibly in isotropic, polycrystalline bodies. However, in the case of Al_2O_3 and Al_2O_3 - 13 w/o TiO_2 plasma-sprayed coatings, it was found that a porosity decrease from 5-6 vol. percent to less than 2% by controlling

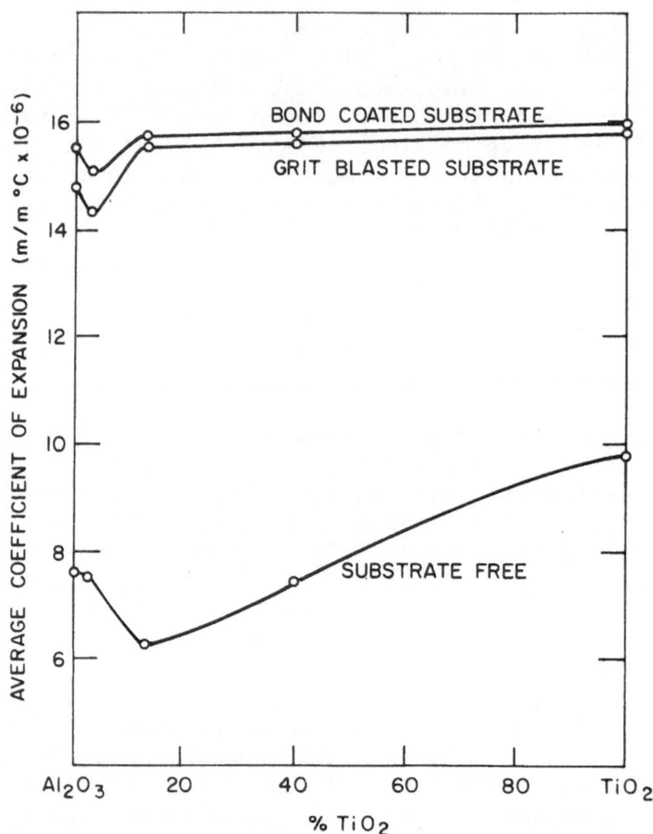

Figure 2: Average thermal expansion coefficient of substrate free and substrate attached Al_2O_3-TiO_2 coatings at 1000°C.

spray parameters resulted in a slight decrease in thermal expansion. This decrease, we thus conclude is likely the result of spray-induced features. Relative to this assumption, Hasselman and Singh observed that an increase in microcracking in brittle heterogeneous ceramics leads to a decreased thermal expansion coefficient(7). It can be assumed that decreased porosity in a coating will lead to an increase in microcracking if the coating is to elongate upon heating. Thus, it follows that a decreased porosity would be expected to give rise to a decreased thermal expansion coefficient through the occurrence of microcracking.

Coatings Attached to Substrate

Expansivity was determined for coatings sprayed onto grit-blasted substrates and metallic, intermediate adhesion-enhancing bond coats (Ref. Fig. 2). The results indicate that coatings attached to the substrate show a substantial increase in the expansivity, approaching the value of the steel substrate. For example, a pure Al_2O_3 coating which is substrate-free has an α value which is about one-half that of the attached Al_2O_3 coating, indicating that the latter is following the expanding substrate. The minimum value of α for the attached coating shifts to lower percentages of TiO_2, this effect being observed for both the grit-blasted and the bond-coated coatings, although the latter coatings showed slightly higher values of . This increase in α is primarily due to the high adhesion between the coating and the substrate that is influencing the expansivity. The higher expansivity reported for coatings with adhesion-enhancing bond coats confirm this observation.

The flame-stabilized zirconia coatings generally displayed greater expansion and, thus, are more compatible with the metal substrate. Zirconia coatings are also less susceptible to failure (e.g., cracking and spallation) during furnace tests. Again, it is generally observed that the thermal expansion coefficient is larger by a factor of 1.5-2.0 for coatings attached to substrates as compared with substrate-free coatings. Intermediate adhesion-enhancing bond coats once again accentuate this effect, indicating that the bond coat is being pulled along with the expanding substrate during both heating and cooling.

Thermal Shock Resistance

To determine the effects of thermal expansion mismatch on thermal shock resistance, candidate coatings of Al_2O_3, Al_2O_3-TiO_2, and ZrO_2 (stabilized by 20 wt. % - Y_2O_3) were furnace cycled between 25°-1000°C at 30 min. cycles in argon sealed tubes. From the results of these thermal fatigue tests, it was evident that the lower expansion materials like Al_2O_3/TiO_2 composites have inferior thermal fatigue resistance as compared with zirconia

Figure 3: Photomicrograph of thermally cycled Al_2O_3

Figure 4: Photomicrograph of thermally cycled Al_2O_3-13% TiO_2.

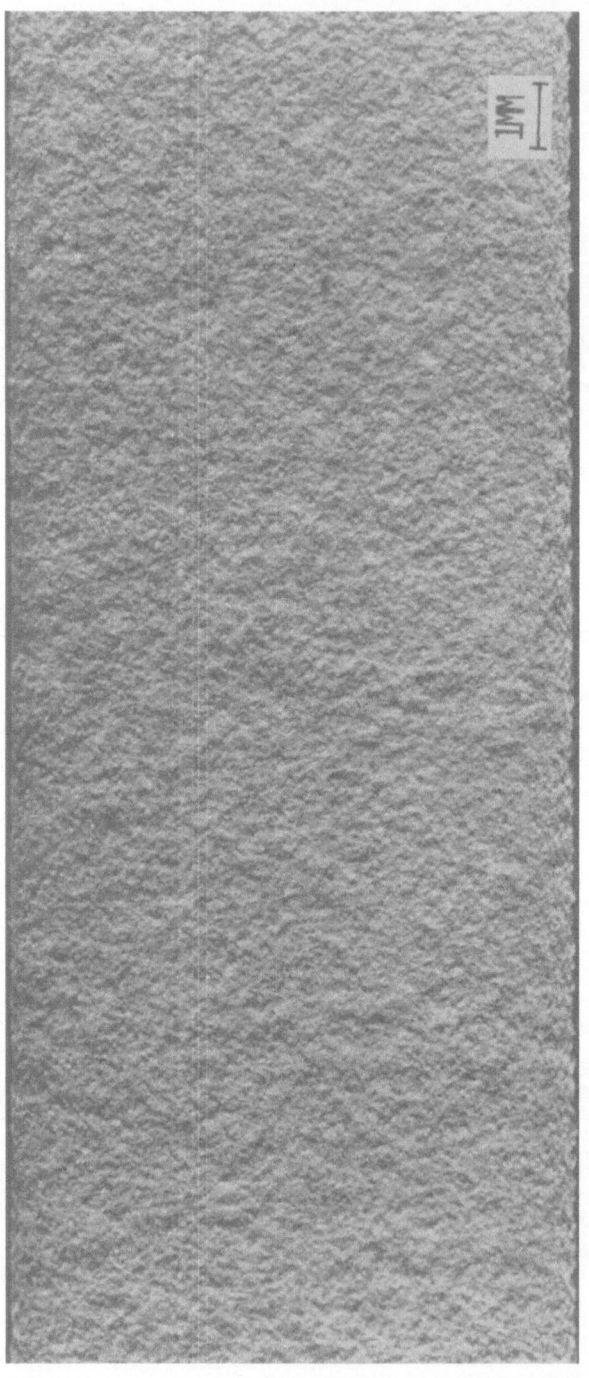

Figure 5: Photomicrograph of thermally cycled ZrO_2 (Y_2O_3).

coatings. Spallation and transverse macro-cracking were observed in Al_2O_3 and Al_2O_3-TiO_2 composites, whereas the zirconia coatings showed microcracking under metallographic examination and did not fail during the duration of this test. The lowest expansion material, Al_2O_3 - 13 w/o-TiO_2, showed random radial macro-cracks, whereas the Al_2O_3 failed by spallation, showing larger and fewer macro-cracks. (Figs. 3-5).

CONCLUSIONS

Using a simple modification of a single silica push-rod dilatometer, it has been possible to evaluate the thermal expansion characteristics of plasma-sprayed oxide coatings. The results of variations on thermal expansion behavior during thermal cycling of plasma-sprayed coatings indicate the importance of this measurement to assess the integrity of coatings, and also, more generally, the importance of dilatometry in the characterization of sprayed coatings. The results of these experiments indicate matching the expansion coefficient of sprayed coatings with that of the substrate is important to the thermal shock resistance of the coatings; a view opposite to that in bulk forms, where minimum expansivity assures better thermal shock resistance.

REFERENCES

1. S. Rangaswamy, H. Herman, S. Safai, Thin Solid Films, 73 (1980) 43.

2. W. R. Bressan, N. K. Thielk, R. V. Sarakankas, Ceramic Age, 60 (5) (1952) 38.

3. O. J. Whittermore, N. N. Ault, Journal of the American Ceramic Society, 39 (1956) 443.

4. S. Safai, "Microstructural Investigation of Plasma Sprayed Metal and Oxide Coatings", Ph.D. Thesis (1979) SUNY at Stony Brook.

5. R. L. Coble, W. D. Kingery, Journal of the American Ceramic Society, 39 (1956) 337.

6. J. B. Austin, Journal of the American Ceramic Society, 14 (1931) 795.

7. D. P. H. Hasselman, J. Singh, American Ceramic Society Bulletin, 58 (1979) 856.

THERMAL EXPANSION OF N.B.S. STANDARD REFERENCE MATERIAL 737 (TUNGSTEN) BELOW 80K

C.A.V. de A. Rodrigues, J. Plusquellec and P. Azou*

Laboratoire Gaz dans les Métaux, Université Paris-Sud, 91405 Orsay, France
*Laboratoire Hydrogène et Matériaux, Ecole Centrale des Arts et Manufactures, 92290 Châtenay-Malabry, France

ABSTRACT

The tube-type vitreous silica vertical differential dilatometer[1], recently interfaced in our laboratory with a micro-computer in order to automatically measure the thermal expansion of solids from 4.2 to 350K[2], was employed to extend N.B.S. thermal expansion measurements on N.B.S. certified tungsten (certified in the 80 to 1800K temperature range) from 80 to 20K using N.B.S. certified copper (SRM 736) as a reference material. A summary of the main features of the apparatus previously described[2] are reported on. Measurements, with calibrated thermocouples, on N.B.S. certified copper and tungsten in the 20 to 300K temperature range which were used to calibrate the apparatus in the 80 to 300K range using a calibration scheme similar to that proposed by Plummer[3] are presented. A least squares polynomial fit was performed to the expansion data on SRM 737 and the expansivity was obtained by taking the derivative of the equation obtained for themal expansion measurements.

INTRODUCTION

For several years we have been working on the development of a tube-type low temperature differential dilatometer for the study of low temperature phase transformations and connected phenomena. Lately, our effort has been devoted to its improvement in order to increase its sensitivity and efficiency and to facilitate its operation. The apparatus described[1] during the 7th ITES in its analog version was therefore recently interfaced with a micro-computer to automatically measure the thermal expansion of solids from 4.2 to 350K[2].

Since it is common for this type of dilatometer to have systematic errors on the order of 100 p.p.m. in measurements of $\Delta L/L$ the calibration of the apparatus was realized using two N.B.S. Standard Reference Materials. The National Bureau of Standards in a program to establish a series of Standard Reference Materials for thermal expansion has certified the following materials using an interferometer technique to an accuracy of about 10 p.p.m.[4] :

SRM 736[5] - copper, certified from 20 to 800K ;
SRM 739[6] - fused silica, certified from 80 to 1000K ;
SRM 731[4] - borosilicate glass, certified from 80 to 680K ;
SRM 737[7] - tungsten, certified from 80 to 1800K ;
SRM 732[7] - sapphire single crystal, certified from 293 to 2000K.

These materials are extremely useful in the testing and calibration of push rod dilatometers allowing for correction of their systematic errors and for evaluation of the accuracy of their thermal expansion measurements. Copper SRM 736 and tungsten SRM 737 were used to calibrate the dilatometer in its digital version from 300 to 80K (lower limit of certification for SRM 737) using an automatic data acquisition system controlled by a micro-computer.

Since SRM 736 was certified to 20K, it was decided to extend measurements which were to be made for the calibration of the apparatus downward to 20K. The expansion data obtained for four runs in the 20 to 300K temperature range were averaged and used to obtain a least squares polynomial fit in temperature. The expansivity was obtained by taking the derivative of the equation for thermal expansion.

APPARATUS

A schematic of the improved version of the apparatus, described in reference 1, is given in Fig.1. The main features of the dilatometer as described in reference 2 are :

- sensitivity to relative length changes for a specimen 50 mm long : $\frac{\Delta L}{L} = 5 \times 10^{-7}$;
- useful range of the LVDT : $\pm$ 1mm ;
- drift rate of the dilatometer at room temperature and at 77K : $< 2.5 \times 10^{-4}$ mm/24 hours ;
- linear cooling and heating rates from 4.2 to 350K : 0.04K/min to 5K/min ;
- temperature stability of the sample and reference during isothermal holding in the 4.2 to 350K temperature range : $\pm$ 0.1K.

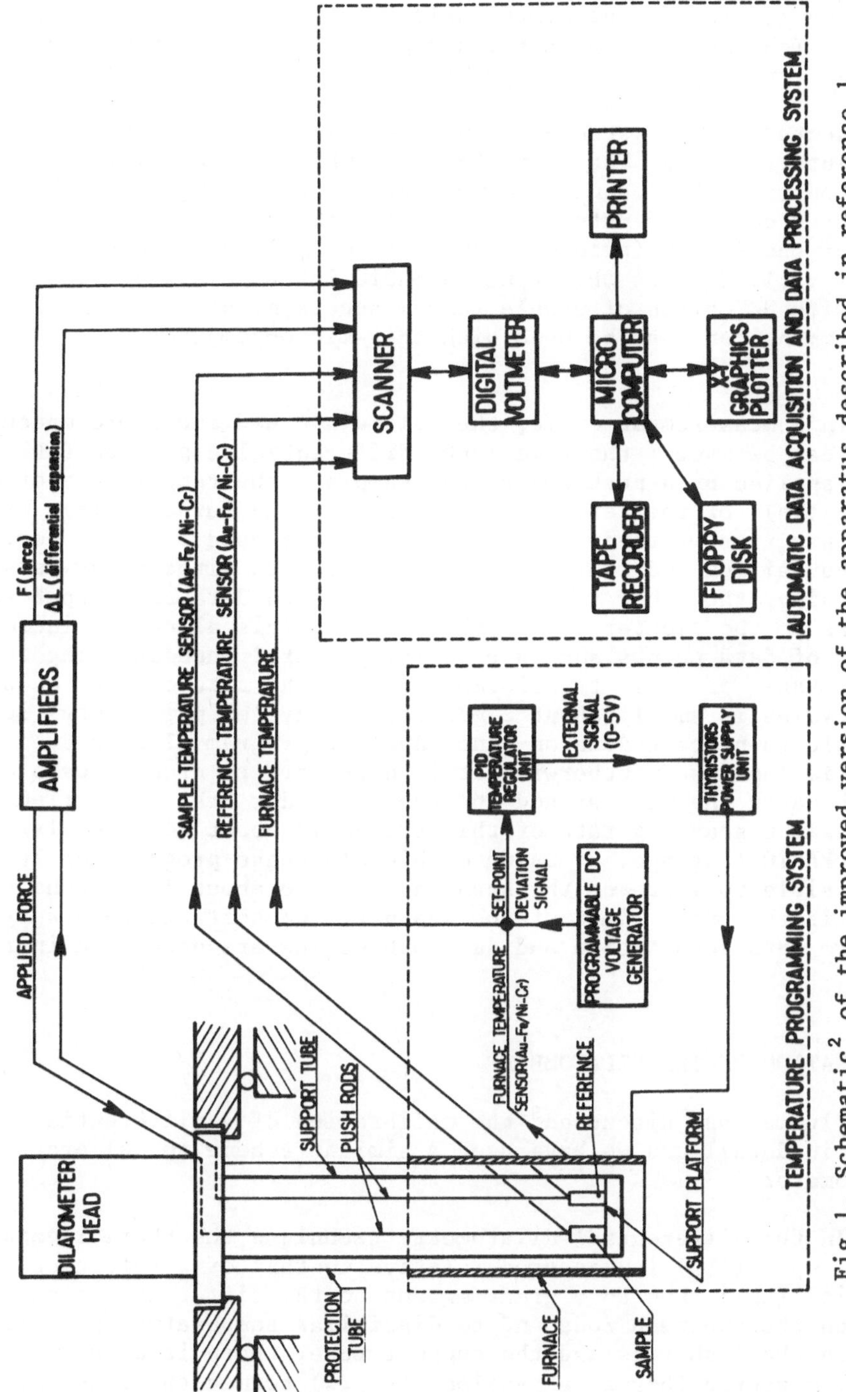

Fig.1. Schematic[2] of the improved version of the apparatus described in reference 1.

As shown in Fig.1, the apparatus is connected to an automatic data acquisition and data processing system which consists of : a scanner (channel multiplexer), a digital voltmeter, a micro-computer (system controller), a line printer, a cassette tape-recorder, a dual drive floppy disk unit, and an X-Y graphics plotter. All the components of the system are connected via the IEEE 488-1975 standard interface bus. The micro-computer which is the system controller, controls the scanner and voltmeter, stores and processes the collected data, and outputs the results to the "external" peripherals of the system (printer, tape-recorder, floppy disk unit, plotter...). It uses the standard BASIC language and provides the user with 32K bytes of usable random access memory. Machine language programming can also be used with this micro-computer.

In a measurement cycle, the following read and store measurements can be made : the time ; the differential expansion (ΔL) ; the force applied by a push rod on the sample ; the temperature of the sample (Ts), of the reference (Tr), and of the furnace (Tf). Two major programs written in BASIC language are used : one to perform the acquisition and storage of transducer data without any data processing, the other to process the acquired data and output the results to the plotter and/or the printer. This allows an increased number of data points and is especially useful when studying the first stages of phase transformations. If the rate of data acquisition cycles is small (about 2 measurement cycles per minute) it is possible to process data on-line. Another program also written in BASIC is then used. Otherwise it is necessary to process data off-line because the plotter and the tape-recorder are slow peripherals. The maximum scanning rate of this system is about 20 channels/s using BASIC language. By using machine language programming it should be possible to increase the scanning rate to about 125 channels/s which is the upper limit of the scanner. The software necessary to perform data acquisition and data processing are described in reference 2.

CALIBRATION OF THE DILATOMETER

Plummer has discussed[3] the calibration of a differential dilatometer in detail and we have used a similar scheme to calibrate the dilatometer.

In the differential dilatometry technique the thermal expansion of a sample (ΔL_s) is measured relative to that of a reference material (ΔL_r). Due to a great extent to the difference in expansion between the two push rods and to dissimilar temperature profiles between the push rods and the support tube, there is a contribution to the measured thermal expansion (ΔL_{meas}) termed the baseline contribution (ΔL_{bl}). This contribution must be determined over the entire temperature range of the apparatus in order to correct the measured expansion. Hence, the measured expansion is of the form :

$$\Delta L_{meas} = C \times (\Delta L_s - \Delta L_r + \Delta L_{bl}) \qquad (1)$$

where C and ΔL_{bl} are calibration constants for the system.

The baseline contribution can be determined by a pair of measurements on two samples (s_1 and s_2) as follows :

$$\Delta L_{meas1} = C \times (\Delta L_{s1} - \Delta L_{s2} + \Delta L_{bl}) \qquad (2)$$

$$\Delta L_{meas2} = C \times (\Delta L_{s2} - \Delta L_{s1} + \Delta L_{bl}) \qquad (3)$$

Adding expressions (2) and (3) gives

$$\Delta L_{bl} = \frac{1}{2xC} \times (\Delta L_{meas1} + \Delta L_{meas2}) \qquad (4)$$

Once the baseline contribution has been determined, the calibration constant C can be obtained by measuring the differential expansion of two reference materials (r1 and r2) and is given by the expression

$$C = \frac{\Delta L_{meas} - C \times \Delta L_{bl}}{\Delta L_{r_1} - \Delta L_{r_2}} \qquad (5)$$

where $C \times \Delta L_{bl}$ is given by (4) and $\Delta L_{r_1} - \Delta L_{r_2}$ can be calculated from tables or equations representing the thermal expansion of the reference materials.

Since the apparatus is intended to be used in the study of low temperature phase transformations and since linear cooling and heating rates of 1 to 2K/min are generally performed, the calibration constants were determined using a cooling and heating rate of 1K/min. An alternative and more accurate procedure was described by Koolie et al.[8]

The temperature was measured by placing the hot junction of one of the two calibrated thermocouples (Au − 0.03 at % Fe and Chromel wires) into the drilled hole of a copper dummy specimen filled with a thermal bonding agent based on silicones and fillers[9]. The dummy specimen was placed on the support tube in the vicinity of the sample and reference position and helium exchange gas was used in the dilatometer chamber. The Au-Fe vs. Chromel thermocouples were calibrated relative to the EPT-76 scale up to 27K and relative to the IPTS-68 scale for temperatures above by Cryogenic Calibrations Ltd. (UK) which reported[10] that temperature measurements may have systematic errors as high as 0.2K over the entire temperature range of the apparatus. Berman et al.[11-14] have made detailed studies on this type of thermocouple and have discussed its usefulness throughout the range 1 - 300K : sensitivity is greater than 10μV/K throughout the above mentioned temperature range ; and the calibration is reproducible after aging and thermal cycling.

The baseline contribution was determined by a pair of measurements on two OFHC copper samples over the 5 to 300K temperature range whereas the calibration constant C was determined in the 80 to 300K range by four repeated measurements in the 20 to 300K temperature range on N.B.S. certified copper (SRM 736) and tungsten (SRM 737). Data points collected at 0.5K temperature intervals during cooling and heating were averaged giving a unique set of thermal expansion data for each measurement. Furthermore, the four sets of thermal expansion data thus obtained on SRM 736 and SRM 737 were averaged and the mean values were substituted in expression (5) at 10K temperature intervals giving 23 values for C as a function of temperature. These were then averaged giving a unique and "best" value for C in the 80 to 300K temperature range.

RESULTS AND DISCUSSION

The averaged data from the four sets of thermal expansion measurements obtained on SRM 737 relative to SRM 736 are given as a function of temperature in Table 1. A least squares polynomial fitting was performed to the experimental data which were found to fit the equation

$$\frac{L_T - L_{293}}{L_{293}} \times 10^6 = -863.31309 - 1.43345 \times T + 3.20084 \times 10^{-2}\, T^2 - 8.30106 \times 10^{-5}\, T^3 + 8.45934 \times 10^{-8}\, T^4 \qquad (6)$$

The expansivity obtained by taking the derivative of the above equation is also given in Table 1 along with N.B.S. certified values of expansion and expansivity.

Figure 2 is a plot of the thermal expansion of SRM 737. A few of our experimental points are shown to indicate the data spread and the smooth curve was obtained with equation (6). Figure 3 is a plot of the difference (in %) between the derivative of equation (6) and the N.B.S. certified values of expansivity and reveals a difference of less than ± 2.30%.

Although the thermal expansion measurements on SRM 737 obtained in the present investigation are in very good agreement with that of the N.B.S. (Table 1) the agreement for expansivity is not as good (Fig.3). Furthermore, our values of expansivity at 65K and below differ significantly with the values of White et al.[15]. This is partly due to the least squares fitting program which is still being developed and improved and should allow us to obtain better results for α.

Table 1. Thermal expansion and expansivity for SRM 737 (tungsten)

Temperature, K	a) $\frac{L_T - L_{293}}{L_{293}}$	a) $\alpha = \frac{1}{L_{293}} \cdot \frac{dL}{dT} (K^{-1})$	b) $\frac{L_T - L_{293}}{L_{293}}$	b) $\alpha = \frac{1}{L_{293}} \cdot \frac{dL}{dT} (K^{-1})$
20	-881×10^{-6}	-0.25×10^{-6} ?		
30	-879	0.27		
40	-874	0.75		
50	-864	1.19		
60	-851	1.58		
70	-833	1.94		
80	-812	2.27	-814×10^{-6}	2.30×10^{-6}
90	-788	2.56	-790	2.61
100	-762	2.82	-762	2.88
110	-733	3.05	-732	3.11
120	-701	3.25	-700	3.30
130	-668	3.42	-666	3.46
140	-630	3.58	-631	3.59
150	-595	3.71	-595	3.71
160	-558	3.82	-557	3.81
170	-519	3.91	-519	3.89
180	-479	3.99	-479	3.97
190	-439	4.06	-439	4.04
200	-398	4.12	-398	4.10
210	-357	4.16	-357	4.15
220	-315	4.20	-315	4.20
230	-274	4.23	-273	4.24
240	-231	4.26	-231	4.27
250	-188	4.29	-188	4.31
260	-145	4.32	-145	4.34
270	-101	4.36	-101	4.36
280	- 57	4.40	- 57	4.39
290	- 13	4.44	- 13	4.41
293	0	4.46	0	4.42
300	31	4.49	31	4.44

a) Present investigation

b) Values for N.B.S. SRM 737 (N.B.S. certificate, 1976).

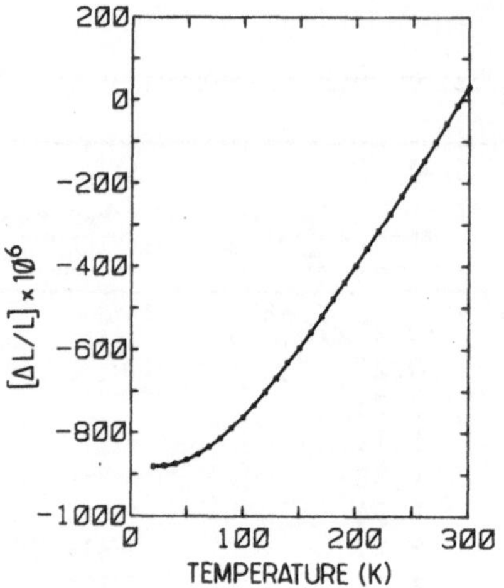

Fig.2. Thermal expansion of SRM 737 ($\frac{\Delta L}{L} = \frac{L_T - L_{293}}{L_{293}}$).

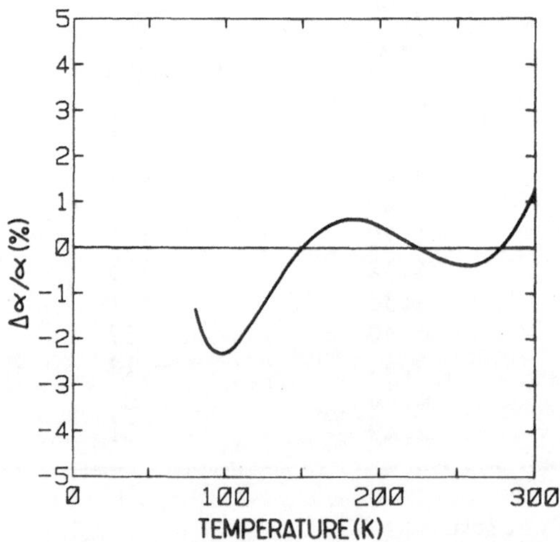

Fig.3. Percent difference between the derivative of equation (6) and the N.B.S. certification of SRM 737 ($\frac{\Delta\alpha}{\alpha} = \frac{\alpha - \alpha_{N.B.S.}}{\alpha_{N.B.S.}}$ x 100).

Nothing has yet been done to check the accuracy of our calibration and should be accomplished by repeated measurements on another set of Standard Reference Materials. Nevertheless, from comparison of the present data with that of the N.B.S., it appears that a reasonable error for the thermal expansion measurements in the 20 to 80K temperature range can be established at about ± 0.30%. Such an evaluation is in relation to what was said about temperature measurements and with the fact that measurements below 80K were realized as an extension of the measurements for the calibration of the dilatometer. Therefore, until more accurate results are obtained we propose our results on SRM 737 as reference data below 80K.

REFERENCES

1. C.A.V. de A. Rodrigues, M. Carrard, J. Plusquellec, and P. Azou, "A Low Temperature (4.2 to 350K) Differential Dilatometer", Thermal Expansion 7, Plenum Press, New York, 67-82 (1982).
2. C.A.V. de A. Rodrigues, J. Plusquellec and P. Azou, "Pilotage par micro-ordinateur de l'acquisition et du traitement des données sur un dilatomètre différentiel très basse température (4,2-350)K", Mém. Sci. Rev. Mét. 79, 149-154 (1982).
3. W.A. Plummer, "Differential Dilatometry, A Powerful Tool", AIP Conf. Proc. No. 17-Thermal Expansion, American Institute of Physics, New York, 147-158 (1974).
4. T.A. Hahn and R.K. Kirby, "Thermal Expansion of a Borosilicate Glass from 80 to 680K - Standard Reference Material 731", AIP Conf. Proc. No. 17 - Thermal Expansion, American Institute of Physics, New York, 93-101 (1974).
5. T.A. Hahn, "Thermal Expansion of Copper from 20 to 800K -Standard Reference Material 736", J. Appl. Phys. 41, 5096-5111 (1970).
6. T.A. Hahn and R.K. Kirby, "Thermal Expansion of Fused Silica from 80 to 1000K - Standard Reference Material 739", AIP Conf. Proc. No. 3 - Thermal Expansion, American Institute of Physics, New York, 13-24 (1972).
7. T.A. Hahn, "Thermal Expansion of Single Crystal Sapphire from 293 to 2000K - Standard Reference Material 732", Thermal Expansion 6, Plenum Press, New York, 191-201 (1978).
8. T.G. Kollie, D.L. McElroy, J.T. Hutton and W.M. Ewing, "A Computer Operated Fused Quartz Differential Dilatometer", AIP Conf. Proc. No. 17 - Thermal Expansion, American Institute of Physics, New York, 129-146 (1974).
9. M.M. Kreitman, "Low Temperature Thermal Conductivity of Several Greases", Rev. Sci. Instrum. 40, 1562-1565 (1969).
10. B.W. Ricketson, Cryogenic Calibrations Ltd. (UK), private correspondence (1980).
11. R. Berman and D.J. Huntley, "Dilute Gold-Iron Alloys as Thermocouple Material for Low Temperature Heat Conductivity Measurements", Cryogenics 3, 70 - 76 (1963).

12. R. Berman, J.C.F. Brock and D.J. Huntley, "Properties of Gold + 0.03 per cent (at.) Iron Thermoelements between 1 and 300K and Behaviour in a Magnetic Field", Cryogenics 4, 233-239 (1964).
13. R. Berman, J.C.F. Brock and D.J. Huntley, "Dilute Gold-Iron Thermoelements in Low - Temperature Thermocouples", Advances in Cryogenic Engineering-Vol. 10, Plenum Press, New York, 233-238 (1965).
14. R. Berman, J. Kopp, G.A. Slack and C.T. Walker, "Magnetic Field Dependence of the Thermoelectric Power for Au + 0.03% Fe at Low Temperatures", Phys. Letters 27A, 464-465 (1968).
15. G.K. White, T.F. Smith and R.H. Carr, "Thermal Expansion of Cr, Mo and W at Low Temperatures", Cryogenics 5, 301-303 (1978).

THERMAL EXPANSION ROUND ROBIN RESULTS

Roger L. Blaine

Du Pont Company - Analytical Instruments Division
Concord Plaza, Quillen Bldg
Wilmington, DE 19898

The American Society for Testing and Materials (ASTM) Committee E-37 on Thermal Measurements is working to produce a Standard Test Method for the measurement of linear thermal expansion using thermomechanical analysis (TMA). This project results from the widespread use of this measurement in the electronics (eg. printed circuit and wiring boards) and aerospace (eg. composite construction materials) industries. The project is assigned the number TM-05-01C and is the responsibility of Task Group E-37.05.01C under the chairmanship of R. L. Blaine

Early in 1979, interlaboratory testing was conducted on polymers, metal, and glass samples using five instrument models to determine the precision and accuracy of the proposed test method. In addition, intralaboratory testing was carried out to determine the precision dependence on heating rate, specimen size, thermal conductivity and coefficient of expansion. This report details the results of those two studies.

Interlaboratory Testing

Interlaboratory testing was carried out on 8 mm specimens of polycarbonate, lead, soft and borosilicate glass. These four materials cover the coefficient of expansion range from 73 to 3 $\mu m/m^{o}C$. An aluminum specimen of the same dimensions was used as the calibration standard. A temperature program rate of $5^{o}C$/minute was used and the measurement made over the range from 0 to $100^{o}C$. The eight instruments participating in the study included Du Pont models 941, 942, and 943, Perkin-Elmer model TMS1 and Theta Dilatronic IV. The data was treated statistically in the manner described in ASTM Method E-180 "Standard Recommended

$$\alpha = \frac{\Delta L}{L\,\Delta T}$$

$$\frac{\delta\alpha}{\alpha} = \frac{\delta\Delta L}{\Delta L} + \frac{\delta L}{L} + \frac{\delta\Delta T}{\Delta T}$$

where:

α = mean coefficient of expansion (μm/m °C)
$\delta\alpha$ = imprecision in α measurement (μm/m °C)
ΔL = change in specimen length (μm)
$\delta\Delta L$ = imprecision in ΔL (μm)
L = specimen length (m)
δL = imprecision in L (m)
ΔT = temperature difference (°C)
$\delta\Delta T$ = imprecision in ΔT (°C)

Fig. 1. Estimation of precision (coefficient of expansion).

$$\frac{\delta\alpha}{\alpha} = \frac{\delta\Delta L}{\Delta L} + \frac{\delta L}{L} + \frac{\delta\Delta T}{\Delta T}$$

where:

ΔL = 60 μm
$\delta\Delta L$ = ± 1 μm
L = 8 mm
δL = ± 25 μm
ΔT = 100°C
$\delta\Delta T$ = ± 0.5°C

$$\frac{\delta\alpha}{\alpha} = \frac{1\,\mu m}{60\,\mu m} + \frac{0.025\,mm}{3\,mm} + \frac{0.5\,°C}{100\,°C} = 0.025$$

$$\frac{\delta\alpha}{\alpha} = 2.5\,\%$$

Fig. 2. Estimation of precision (example).

Practice for Developing Precision Data on ASTM Methods for Analysis and Testing of Industrial Chemicals".

Single instrument repeatability was determined from duplicate determinations. As might be expected, the mean standard deviation (calculated on an absolute basis) decreased from 0.89 to 0.22 μm/m^{o}C in going from coefficient values from 73 to 3 μm/m^{o}C. Correspondingly, the mean coefficient of variation increased from 1.2 to 7.2% for the same range. The pooled single instrument repeatability (standard deviation) for all four samples and eight instruments (ie. 38 degrees of freedom) was 0.97 μm/m^{o}C. Two results should be considered suspect, therefore, if they differ by more than 2.7 μm/m^{o}C.

Interlaboratory reproducibility was determined from mean values of duplicate determinations. The pooled standard deviation for eight instruments was 1.1 μm/m^{o}C. Two results (each the mean of duplicates) should be considered suspect if they differ by more than 3.1 μm/m^{o}C.

To determine the accuracy of the test method, the coefficient value obtained for lead 30.9 μm/m^{o}C was compared with its literature value of 29.33 μm/m^{o}C yielding a variance of 4.5% (rel.) (1)

Intralaboratory Testing

The decreasing standard deviation and increasing coefficient of variation with decreasing expansion observed in the interlaboratory testing suggests the overall experimental error is dependent upon some parameter(s) in addition to the coefficient of expansion itself. Additionally, it was necessary to have information on the effect of changes in experimental conditions held constant during the interlaboratory testing, including specimen size and heating rate. These questions dictated intralaboratory testing.

Intralaboratory testing was carried out using a Du Pont 990 Thermal Analyzer and 943 Thermomechanical Analyzer. Specimen sizes and heating rates were used in addition to those in the interlaboratory testing. NBS coefficient of expansion Standard Reference Material copper was included to get a second estimation of accuracy.

The first step was to examine the equation for calculation of the coefficient of expansion for propagation of uncertainties (2). Figure 1 illustrates that the estimated precision equals the sums of the individual measured component precisions. Typical values taken from the test method substituted into this equation (figure 2) lead to an estimated precision of ±2.5% (rel). This is in excellent agreement with those values obtained in the inter-

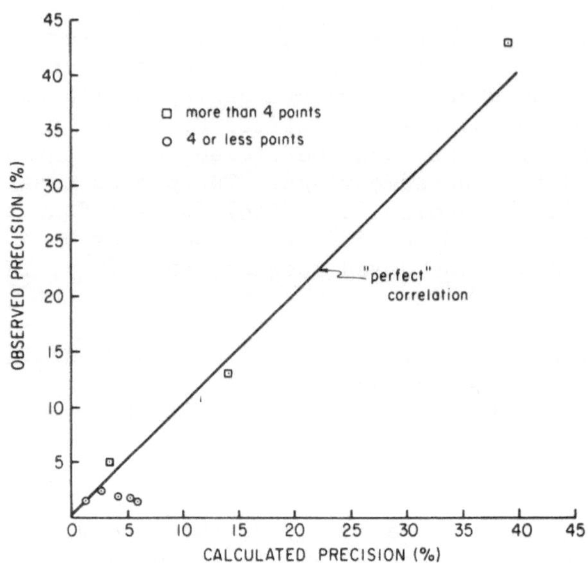

Fig. 3. Precision of expansion coefficient measurement.

property	high	low
program rate (°C/min)	10	2.0
size (mm)	8	2
thermal conductivity (W/cm°C)	4.0	0.0023
coefficient of expansion (μm/m°C)	73	8.3

Fig. 4. Factorial design parameters.

laboratory testing and those of previous authors (3,4). Testing of additional samples lead to the correlation illustrated in figure 3. The observed precision corresponds quite closely to the calculated precision using the equation in figure 1. Thus, the propagation of uncertainties method is judged to be an excellent method to estimate repeatability of an unknown material.

In studying the effects on precision of program rate, specimen size, thermal conductivity and coefficient of expansion, a factorial design was used with the ranges shown in figure 4. These were judged to cover the "normal" experimental range and to be sufficiently separated to show the desired interactions. Since thermal conductivity and coefficient of expansion could not be varied independently, the data was evaluated based on two factorial experiments schematically shown in figures 5 and 6. Thermal conductivity and coefficient of expansion are truly independent since their orders are copper > polycarbonate > glass and polycarbonate > copper > glass, respectively.

The data was evaluated by the "factor effects" method using a 95% confidence limit criterion. The results summarized in figure 7 show no dependence of precision on program rate, specimen size, or thermal conductivity. That is, if only one of these parameters is changed at a time, no effect on precision is observed. This is encouraging since for the testing of a particular type of material, specimen size is the most likely parameter to change. The results do show, however, a slight dependence of precision on coefficient of expansion.

In the case where two parameters are changed from one experiment to the next, no dependence was observed for the important combinations of specimen size - thermal conductivity and size - program rate. If, however, both program rate and thermal conductivity are changed, a small effect is seen in the precision. This effect is quite small, however, since if three parameters are changed at the same time, size, program rate, and thermal conductivity, no effect is seen on the precision.

As a second estimation of the accuracy of the method, the coefficient values for copper obtained in the intralaboratory testing 17.8 μm/m^{o}C were compared with the NBS value of 16.9 μm/m^{o}C for a variance of 5.3% (5). This is in full agreement with the accuracy determined in the interlaboratory work above.

Summary

Inter- and intra- laboratory testing indicate that TMA measurements of the coefficient of expansion can be made with repeatability and reproducibility of 0.97 and 1.1 μm/m^{o}C,

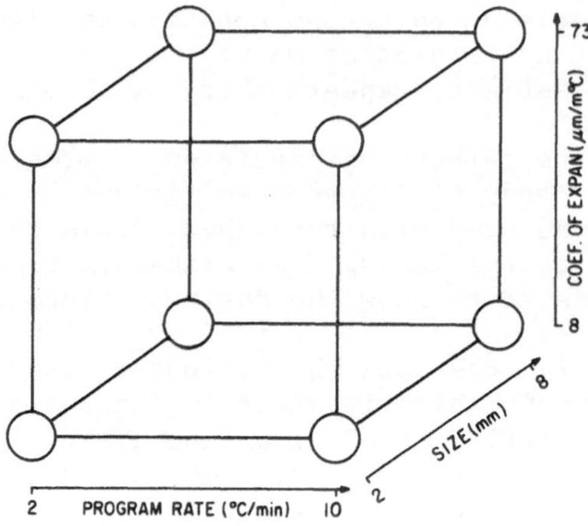

Fig. 5. Factorial design (program rate, size, and coefficient of expansion).

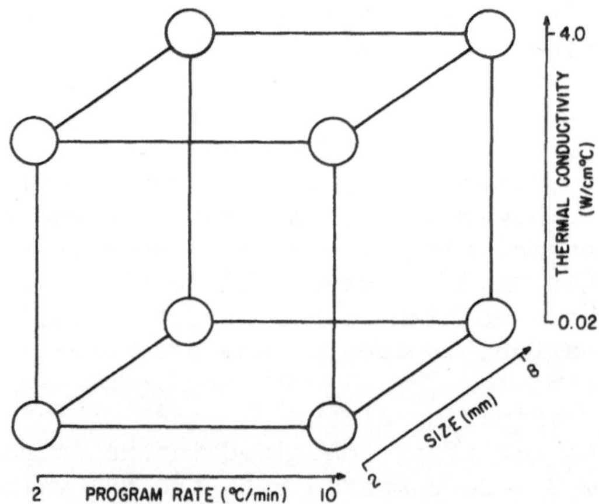

Fig. 6. Factorial design (program rate, size, and thermal conductivity).

SLIGHT DEPENDENCE ON :

- coefficient of expansion
- program rate and sample thermal conductivity (comb.)

NO DEPENDENCE ON:

- program rate (alone)
- sample size (alone)
- sample thermal conductivity (alone)
- program rate, sample size, and thermal conductivity (comb.)

Fig. 7. Analytical precision.

respectively. The method is accurate to ±5.3% (rel). In addition, the precision can be successfully estimated by propagation of uncertainties and is quite independent of the most commonly encountered experimental variables such as sample size and thermal conductivity.

References

1. A. R. Stokes and A. J. C. Wilson, The Thermal Expansion of Lead from $0^{\circ}C$ to $320^{\circ}C$, Proc. Phys. Soc. (London), 53: 658 (1941)
2. D. C. Baird, "Experimentation", pp 48, Prentice Hall, 1962.
3. R. Gaskill and E. M. Barrall, Calibration, Precision, and Accuracy of a Du Pont Thermomechanical Analyzer, Thermochim Acta, 12:102 (1975)
4. J. M. Barton, Calibration of a Thermomechanical Analyzer, Thermochim Acta, 29:188 (1979)
5. T. A. Hahn, Thermal Expansion of Copper from 20 to 800K, J. Appl. Phys., 41:5096 (1970)

A THERMAL EXPANSION MEASUREMENT USING PARALLEL SPRING MOVEMENT

Masahiro Okaji, Hidetaka Imai, Namiteru Hida, and Kozo Iizuka

National Research Laboratory of Metrology
1-1-4, Umezono, Sakura-Mura, Niihari-Gun
Ibaraki, 305, Japan

INTRODUCTION

A thermal expansion measurement system with a unique specimen holder is described. This system is an application of parallel spring component together with a He-Ne laser interferometer. The measurement using this system minimizes the restriction based on specimens.

The most familiar absolute measuring method of thermal expansion is an interferometric dilatometer. However, in the case of the usual interferometry, such as Fizeau method, it consists of two parallel transparent plates on both sides of a specimen, so that parallelism of the specimen ends is strictly required. The system described here minimizes the restrictions on shapes and dimensions of the specimens.

The method consists of the combination of an optical heterodyne interferometer and a specimen holder. The merit in using an optical heterodyne method is that parallelism between two optical paths becomes less critical than in using a usual interferometry.

A combined parallel spring component is used for supporting the system of the specimen and mirrors. The contact between the specimen and two mirrors can be maintained by the parallel spring component. In this system, the expansion or contraction of a specimen caused by a temperature change is transmitted to the parallel spring component to which the optical mirrors are connected. The specimen holder keeps the parallelism between the two mirrors in place of the specimen itself, and specimens of various shapes and dimensions can be measured.

In order to confirm the accuracy of the system, three kinds of Standard Reference Materials, supplied by the National Bureau of Standards, USA, and a fused silica cylinder, and a thin steel plate were measured. The results and discussions of the measurements in the temperature range of 10°C to 60°C are presented.

APPARATUS

An optical design of the dilatometer is shown in Figure 1 (Imai and Bates, 1981). A two-frequency laser (Hewlett Packard Co., model 5500C) and a laser display (HP, model 5505A) are used as a part of the interferometry system. The light coming from the laser head has two components with frequencies f_1 and f_2. The f_1 component is linearly polarized in the plane of the illustration, and f_2 component is perpendicular to the plane of the f_1 component. Upon reaching the Wallaston prism, the two components are split symmetrically with each other, and the two beams are then made mutually parallel by the bi-prism. After each beam is reflected by the two mirrors located at both ends of the specimen, they are recombined by the same bi-prism and Wallaston prism, and finally received by the photo-detector. This recombined light contains relative phase information related to the difference between the two optical paths. The change in interval between the two mirrors can be detected by comparing this signal with the reference signal. The optical arrangement of the system can be finely aligned by using the rolling and tilt adjusters of the Wallaston prism and bi-prism which are placed outside the oven.

Figure 2 shows the mechanical arrangement of the specimen holder. By using bent leaf springs, the front reflector is fastened to the block which is rigidly set up to the base plate. The lower platform of the parallel spring component is rigidly fixed to the base plate, and the back reflector is fastened to the upper moving platform of the parallel spring component. The face of

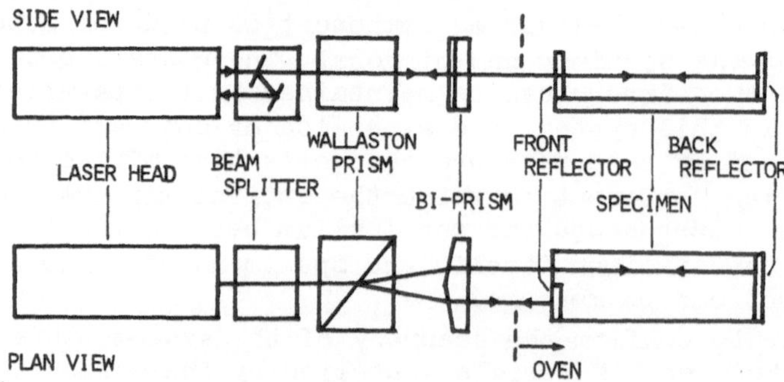

Fig. 1. Optical design.

each mirror is designed to coincide with the contact point of the specimen respectively, within an error of 20 micrometers. The specimen is supported by the force of the parallel spring. As the applied compression or tension stress is very small, the change in length of the specimen due to the change of Young's modulus is negligibly small. In this system, the change in distance between the two mirrors depends only on the length change of the specimen which is transmitted to the movement of the parallel spring component, so that the length change of the base plate does not affect the measurement.

Each part of the specimen holder, except two beryllium copper springs, is made of brass. The difference in the thermal expansion coefficient of the beryllium copper and brass is not so large as to yield any extra thermal distortion. Figure 3 shows a picture of the specimen holder.

The oven contains an aluminum alloy heat sink, and the specimen and the specimen holder are placed in the heat sink. The temperature variation in the specimen is within ± 5mK, when the oven is maintained at constant temperature. The temperature of the inner wall of the heat sink, the specimen, and the specimen holder are measured by several sets of copper-constantan thermocouple. As the measurement is done in air, the measured value of the length change is always corrected with the change in refractive index of the air.

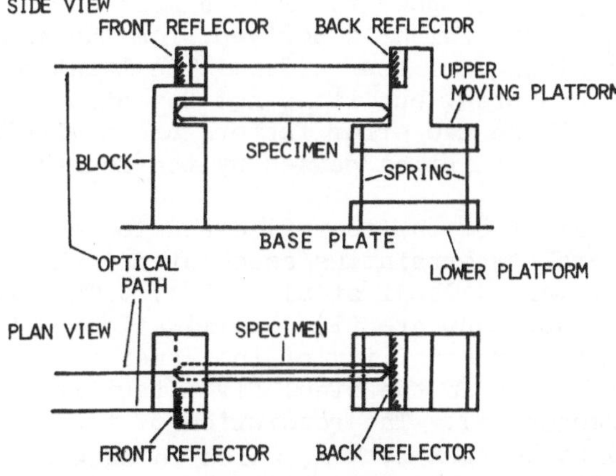

Fig. 2. Mechanical arrangement of the specimen holder.

Fig. 3. Photograph of the specimen holder.

ERROR ANALYSIS

The most important point in this system is how to maintain the parallelism of the parallel spring movement. The error concerning with it is mainly caused by the non-parallelism of the spring movement.

The difference of the thermal expansion coefficient of a specimen and a base plate yields the displacement of the spring component. Non-parallelism of the spring movement is measured by using the same interferometric system. The error caused by the tilt or twist of the back reflector is measured as a function of the displacement of the spring component. As is shown in Figure 4, the tilt ratio is measured to 7×10^{-4}, and this corresponds to 0.07% of the displacement of the spring component. This almost agrees with the calculated value, 0.06%, which is derived from the simple calculation based on the linkage of four sides. The twist of the back reflector is also measured, but it is not detected in the parallel spring component. These two error factors are negligibly small compared with the uncertainties caused by the length and temperature measurements.

In the measurement of the thermal expansion coefficient, there are generally several uncertainties associated with the length and temperature measurements (Okaji et al., 1981). The principal uncertainties in the measurement are (i) determination of the fringe fraction, (ii) long term stability of the laser wavelength, (iii) temperature distribution in the oven, (iv) stability of the thermometer (C-C thermocouple). The resolution of the length measurement is estimated to be $\pm 1 \times 10^{-8}$ m, which is obtained from the measurement of the fringe counting stability over several hours. The long term stability of the laser wavelength is the order of 1 x

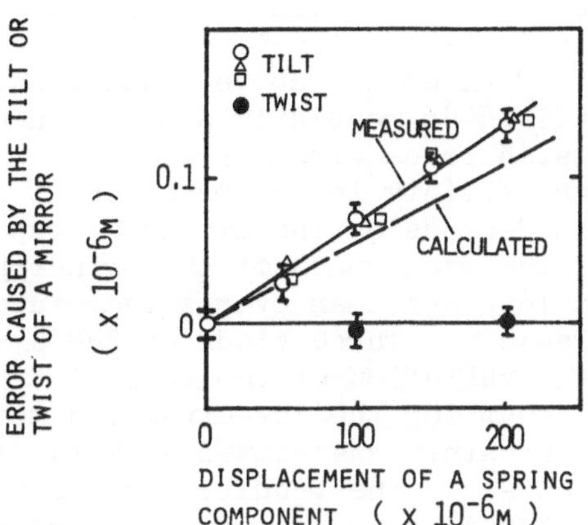

Fig. 4. Measurements of the error caused by the tilt or twist of the back reflector.

Table. 1. Estimated uncertainty (%) associated with the length and temperature measurements

Thermal expansion coefficient (K^{-1})	16×10^{-6}	4.4×10^{-6}	0.5×10^{-6}
Source of uncertainty			
Fringe fraction	0.08	0.9	2.7
Laser wavelength	0.006	0.02	0.2
Temperature distribution	0.4	0.4	0.4
Calibration	0.2	0.2	0.2
Total**	0.5	1.0*	2.7

* calculated for a 52 mm long specimen.
** square root sum.

10^{-9} (Iwasaki and Sakurai, 1980). The measurements are made with temperature intervals of approximately 10°C, and typical time interval for a single measurement is about four to six hours. The temperature distribution in the oven is measured by several sets of C-C thermocouple, and it is confirmed that the variation does not exceed $\pm 2 \times 10^{-2}$ °C under the above conditions. The thermocouples have been calibrated regularly and the uncertainty in the calibration is estimated to be within ± 0.2%.

The estimated uncertainty in the measurement of the thermal expansion coefficient depends on the amount of expansion coefficient itself. Then, the magnitudes of the principal uncertainties for a 152 (or 51) mm long specimen over a temperature interval of 10°C are listed in Table 1. Three kinds of SRM's have been chosen as the examples of the variation of the thermal expansion coefficients. It is clear that for the low-expansion material, i.e. fused-silica, the uncertainty associated with the fringe determination mainly contributes to the result. On the other hand, for the high-expansion material, i.e. copper, the uncertainty associated with the temperature distribution in the oven mainly contributes to the result.

RESULTS

Measurements have been made on five kinds of materials, that is, three Standard Reference Materials (Copper, SRM736L3; Tungsten, SRM737; Fused-silica, SRM739L3), a fused-silica cylinder, and a thin steel plate in the temperature range of 10°C to 60°C.

The measured results of copper, SRM736L3, supplied as a rod of 6.4 mm diameter by 152 mm long, are plotted in Figure 5 against the degree of Celsius scale. By using the specimen holder, the specimen can be measured in its original form, that is 152 mm long, which improves relative resolution of length measurement. This is one of the greatest advantages in this method.

The vertical error bar represents the total uncertainty mentioned in the previous section, and its value is estimated to about $\pm 8 \times 10^{-8} K^{-1}$. The value of the thermal expansion coefficient at 20°C is $16.53 \times 10^{-6} K^{-1}$ and its 95% confidence limit calculated from the standard deviation of the data, $6 \times 10^{-8} K^{-1}$, is almost equal to the estimated total uncertainty.

As is shown in Figure 5, the published results for the thermal expansion coefficient of SRM736 (e.g. Kirby and Hahn, 1975; Kollie et al., 1974; Bennett, 1978) show variation of $0.25 \times 10^{-6} K^{-1}$ (± 0.7% of the thermal expansion coefficient). The result by Bennett seems to have a deviation from the others, but the lowest measured point in his measurement is about 43°C and the regression curve is drawn by extrapolation from the higher temperature range. So, considering the uncertainty of each measured result, these four independent results are in good agreement with each other in the

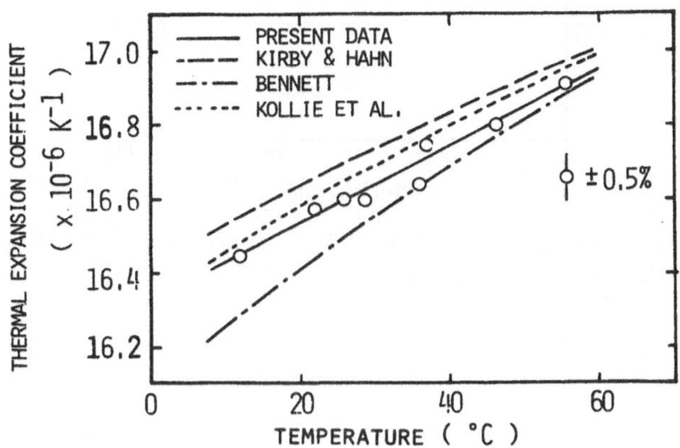

Fig. 5. Measurements of the thermal expansion coefficient of SRM 736, copper.

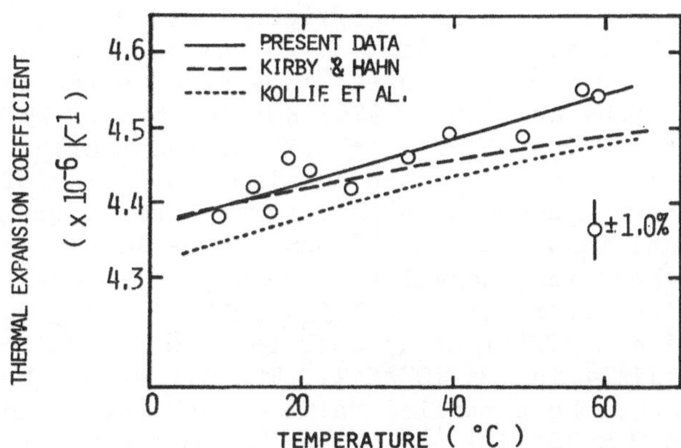

Fig. 6. Measurements of the thermal expansion coefficient of SRM 737, tungsten.

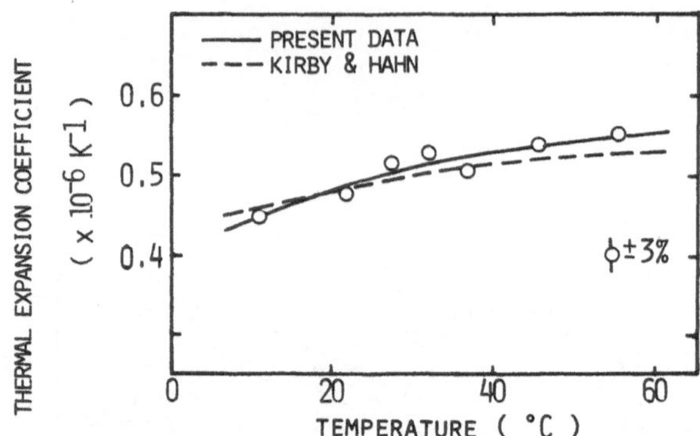

Fig. 7. Measurements of the thermal expansion coefficient of SRM 739, fused-silica.

temperature range of 10°C to 60°C, although the measuring method of each result is different.

Figure 6 shows the thermal expansion coefficient of SRM 737. This tungsten specimen is supplied as a rod of 51 mm long. The amount of the error bar of this specimen is ± 4 x $10^{-8}K^{-1}$. The value of the thermal expansion coefficient at 20°C is 4.42 x 10^{-6} K^{-1}, and its 95% confidence limit is calculated to 3 x $10^{-8}K^{-1}$. As is shown in Figure 6, the three kinds of measured results (present work; Kirby and Hahn, 1976; Kollie et al., 1974) agree with each other within ± 3 x $10^{-8}K^{-1}$ (± 0.7% of the thermal expansion coefficient). It is seen that the maximum deviation of the present result from others appears at higher temperatures, but this difference is not significant at the 95% confidence level.

Figure 7 shows the thermal expansion coefficient of SRM739L3. The amount of the error bar is ± 1.5 x $10^{-8}K^{-1}$. The value of the thermal expansion coefficient at 20°C is 0.48 x $10^{-6}K^{-1}$, and its 95% confidence limit is 2 x $10^{-8}K^{-1}$. The difference between the present result and the certified value of NBS (Kirby and Hahn, 1971) is within 2 x $10^{-8}K^{-1}$ (4% of the thermal expansion coefficient).

These three results show that the measurements of three SRM's by using the present measurement system are in good agreement with other results. As the examples of the measurements of other materials, the results of a fused-silica cylinder and a thin steel plate are shown.

The type-I fused-silica is prepared in the form of 75 mm diameter and 3 mm thick hollow cylinder by 60 mm long. The specimen is in contact with the holder at two points, and the thermal expansion between these two points is measured. The thermal expansion coefficient of the fused-silica cylinder is obtained as 0.39 ± 0.02 (x $10^{-6}K^{-1}$) at 20°C.

In the case of the measurement of the 0.3 mm thick thin steel plate, each end of the specimen is connected with the holder by using a hole of the specimen. And, the specimen is stretched by applying a slight tension. The thermal expansion coefficient of the steel plate is obtained as 11.10 ± 0.05 (x $10^{-6}K^{-1}$) at 20°C.

Both of the above results are in good agreement with those in other literatures, respectively.

CONCLUSIONS

In the present thermal expansion measuring system, an optical heterodyne interferometry is used. In addition, a combined parallel spring component has been developed as a holder of a specimen and two mirrors. The combination of the specimen holder with the interferometer is successful to minimize the restrictions on shapes and dimensions of specimens.

Specimens of three kinds of SRM's, a fused-silica cylinder, and a thin steel plate are measured in the temperature range of 10°C to 60°C with the accuracy of the order of 2 to 8 x $10^{-8}K^{-1}$. The system enables us to measure thermal expansion coefficients of wide variety of specimens with high accuracies, and is being developed for the measurements at higher temperatures.

REFERENCES

1. Bennett, S.J., The Thermal Expansion of Copper Between 300 and 700K, J. Phys. D: Appl. Phys., 11:777 (1978).

2. Imai, H., and Bates, W.J., Measurement of the Linear Thermal Expansion Coefficient of Thin Specimens, to be published in J. Phys. E: Sci. Instr., 14-7 (1981).

3. Iwasaki, S., and Sakurai, T., A Wavelength Calibration of Commercial Wavelength Stabilized He-Ne Laser, Ohyo Butsuri (in Japanese), 49:870 (1980).

4. Kirby, R.K., and Hahn, T.A., NBS Certificate, Standard Reference Material 736 Copper - Thermal Expansion (1975).

5. Kirby, R.K., and Hahn, T.A., NBS Certificate, Standard Reference Material 737 Tungsten - Thermal Expansion (1976).

6. Kirby, R.K., and Hahn, T.A., NBS Certificate, Standard Reference Material 739 Fused Silica - Thermal Expansion (1971).

7. Kollie, T.G., McElroy, D.L., Hutton, J.T., and Ewing, W.M., A Computer Operated Fused Quartz Differential Dilatometer, AIP Conf. Proc., No. 17 - Thermal Expansion, p. 129, AIP, New York (1974).

8. Okaji, M., Imai, H., Hida, N., and Iizuka, K., An Interferometric Dilatometer and Supporting Systems of the Specimen, Ohyo Butsuri (in Japanese), 50:714 (1981).

LOW TEMPERATURE THERMAL EXPANSION OF ZnF_2

G. Diver and A.S. Pavlovic

West Virginia University

Morgantown, WV 26506

INTRODUCTION

Thermal expansion measurements can yield valuable information about the thermodynamic state and lattice properties of the material under study. Although the thermal expansion of many cubic systems has been extensively studied, relatively few noncubic single crystal systems have been measured throughout their complete solid state phase. ZnF_2 is a noncubic material whose thermal expansion from room temperature to 770 K was measured by Rao and Naidu[1] using x-ray diffraction methods. In this paper the results of a detailed thermal expansion study of single crystal zinc difluoride from room temperature down to 53 K are presented. The temperature dependence of the thermal expansion, the coefficients of linear thermal expansion and the principal Grüneisen components were evaluated for the two principal crystal directions. In addition the coefficient of volume thermal expansion was determined and was used in conjunction with specific heat data in order to examine the temperature dependence of the Grüneisen parameter. At room temperature it was found that $\alpha_{\parallel} = 11.5 \times 10^{-6}K^{-1}$, $\alpha_{\perp} = 9.7 \times 10^{-6}K^{-1}$ and $\beta = 31 \times 10^{-6}K^{-1}$ which are all in good agreement with previously reported results.

Zinc fluoride is one of the iron group difluorides which crystallizes in the rutile structure. It has a body-centered tetragonal structure represented by the space group D_{4h}^{14} with Zn^{++} ions located at the unit cell corners and at the body center position. There are two zinc ions per unit cell. Each zinc ion is surrounded by fluorine ions situated at the vertices of a distorted octahedron. Zinc ions form the corners of isosceles triangles which surround and are coplanar with each fluorine ion. Stout and Reed[2] measured the room temperature lattice constants to be a = 4.7034 Å and c = 3.1335 Å, yielding a c/a ratio of 0.6662 and a mass density of 4.950 grams/cm^3. Investigations done on ZnF_2 are important due to the fact that it is the diamagnetic member of the isomorphous series of iron group difluorides. The other members are all antiferromagnetic and thus have magnetic contributions to their measured properties at temperatures near and below their Neél temperatures.

EXPERIMENTAL PROCEDURE

The samples used in this study were cut from a cylindrical single crystal of optical quality which was kindly provided by Dr. R. Fiegelson of Stanford University. The individual samples used varied in size from 6 x 10 x 2 mm to 3 x 5 x 2 mm. The sample crystals were oriented to within ±.5° of the exact orientations using back reflection Laue x-ray patterns.

The measurements were made using standard strain gage techniques. Strain gages were attached to the samples with epoxy along the directions to be measured (i.e. along the a- and c-directions). The thermal expansion measurements were taken using a d.c. Wheatstone bridge circuit. The data was corrected for the thermal expansion of a high purity copper dummy. Several samples were used to collect the data. The spread in the results is about 4%. The samples were held in the sample holder by a spring arrangement that provided ample freedom to dilate.

A pressure regulator was used to stabilize the temperature below 77 K. Above 77 K, the temperature was controlled by using an Artronix model 5301D heat-temperature controller. All measurements were made at a temperature stabilization of ±0.05 K as determined by a copper-constantan thermocouple. Below 77 K, data was taken every 1 K; while above 77 K, data was taken approximately at 7 K intervals.

RESULTS AND DISCUSSION

In Figure 1 are presented representative thermal expansion data from measurements for the a- and c-directions of the single crystals from 53 K to room temperature. The fitted curves were obtained using a least squares fit of a third order polynomial. In both directions the strain decreases in a monotonic fashion as the temperature is decreased. To within the experimental errors and data fluctuations the thermal expansion in the c-direction essentially decreases to a constant around 70 K. On the other hand, the a-direction thermal expansion is still decreasing at 53 K. The least squares curve for the c-direction thermal expansion

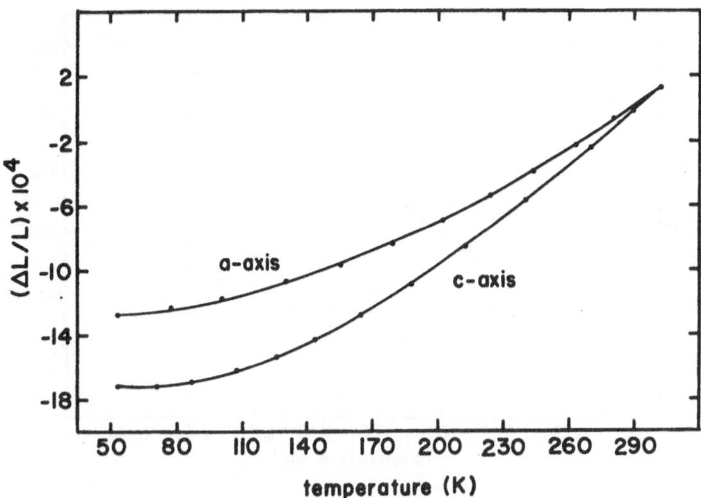

Fig. 1. Thermal expansion along the a- and c-directions of ZnF_2 as a function of temperature.

is represented by

$$\left(\frac{\Delta\ell}{\ell}\right)_c \times 10^6 = -1.53 \times 10^3 - 6.41\ T + 5.89 \times 10^{-2} T^2 - 6.45 \times 10^{-5} T^3,$$

while in the a-direction

$$\left(\frac{\Delta\ell}{\ell}\right)_a \times 10^6 = -1.27 \times 10^3 - 1.02\ T + 2.07 \times 10^{-2} T^2 - 6.02 \times 10^{-6} T^3.$$

The coefficient of linear thermal expansion $\alpha(T)$ is defined by $\alpha(T) = (1/\ell)(d\ell/dT)$. The coefficients of linear thermal expansion for ZnF_2 were obtained by differentiating the above equations. The coefficients of linear expansion along the a-direction ($\alpha_a = \alpha_\perp$) and along the c-direction ($\alpha_c = \alpha_\parallel$) are shown in Figure 2. Also included is a curve representing the powdered (polycrystalline) value of $\alpha(\alpha_p = 2/3\ \alpha_\perp + 1/3\ \alpha_\parallel)$. All three curves decrease smoothly throughout the temperature range examined with $\alpha_\parallel$ effectively equal to zero below 60 K. The least squares curve for $\alpha_\parallel$ had to be truncated below 71 K because the fitting introduced spurious values due to the flatness of $(\Delta\ell/\ell)_c$ below 70 K. At room tem-

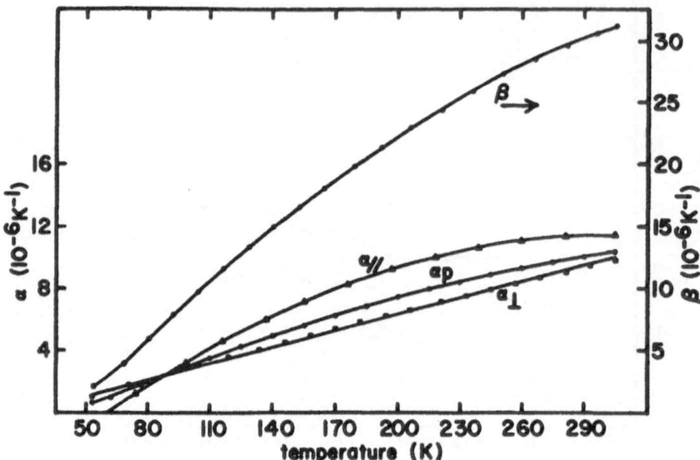

Fig. 2. The coefficients of linear thermal expansion along the a-axis and c-axis and the coefficient of volume thermal expansion for ZnF_2.

perature $\alpha_{\parallel} = 11.5 \times 10^{-6}K^{-1}$ and $\alpha_{\perp} = 9.7 \times 10^{-6}K^{-1}$ which compare well with the values of $\alpha_{\parallel} = 11.6 \times 10^{-6}K^{-1}$ and $\alpha_{\perp} = 8.7 \times 10^{-6}K^{-1}$ obtained by Rao and Naidu[1,3] in their high temperature study of the thermal expansion of rutile structure materials.

The volume coefficient of thermal expansion for tetragonal crystals is given by $\beta = 2\alpha_{\perp} + \alpha_{\parallel}$. The temperature dependence of β for ZnF_2 is also shown in Figure 2. The form of the curve is very similar to that obtained by Browder[4] in his thermal expansion study of MgF_2. At room temperature a value of $31 \times 10^{-6}K^{-1}$ was obtained for the coefficient of volume thermal expansion which is in good agreement with the value of $29 \times 10^{-6}K^{-1}$ obtained by Haefner.[5]

In order to interpret the thermal expansion data the Grüneisen model is usually adopted. The coefficients of linear thermal expansion are used to calculate the Grüneisen parameter $\gamma(T)$ given by $\gamma(T) = \beta V/C_p\chi_s$ where β is the coefficient of volume thermal expansion, V is the molar volume, C_p is the molar heat capacity at constant pressure, and χ_s is the adiabatic compressibility. For ZnF_2 the Grüneisen parameter can be written as

$$\gamma=(\alpha_{\parallel} + 2\alpha_{\perp})V/C_p\chi_s.$$

In order to describe the directional behavior of an anisotropic solid, the Grüneisen tensor is needed. This is because the dilation of an anisotropic solid is a function of the individual strain components and thus the expansion coefficients are tensor quantities. For crystals with tetragonal symmetry there are three components of the Grüneisen tensor, viz. γ_1, γ_2, and γ_3 as given by

$$\gamma_1 = \gamma_2 = [(C_{11} + C_{12})\alpha_{\perp} + C_{13}\alpha_{\parallel}]V/C_p$$

and

$$\gamma_3 = (2C_{13}\alpha_{\perp} + C_{33}\alpha_{\parallel})V/C_p$$

where C_{11}, C_{12}, C_{13}, C_{33} are the adiabatic elastic constants, V is the molar volume, and C_p is the molar heat capacity at constant pressure.

Figure 3 displays the results of the determination of $\gamma(T)$, $\gamma_1 = \gamma_2$, and γ_3. The adiabatic elastic constants used in the calcu-

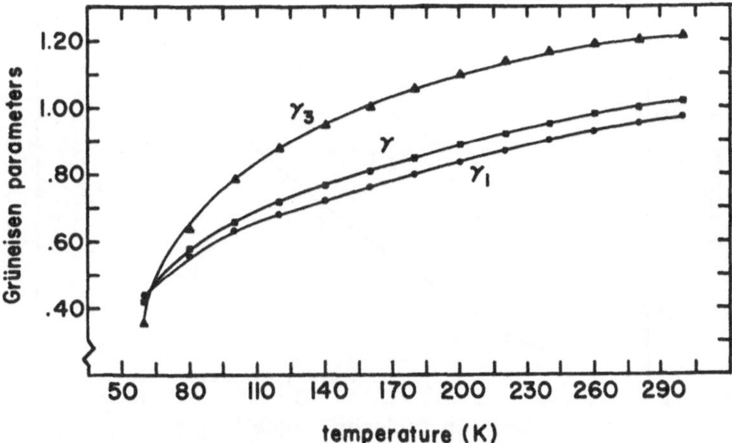

Fig. 3. The Grüneisen parameter and the principal components of the Grüneisen tensor for ZnF_2.

lations were taken from Rimai.[6] The adiabatic compressibility was determined from the adiabatic elastic constant data by using the expression

$$\chi_s = \frac{C_{11} + C_{12} + 2C_{33} - 4C_{13}}{(C_{11} + C_{12})C_{33} - 2C_{13}^2} \ .$$

The temperature dependence of the molar volume was obtained by using the volume coefficient of thermal expansion. Finally, the heat capacity at constant pressure for ZnF_2, obtained by Stout and Catalano[7] was used in the evaluation of the various γ's.

From Figure 3 it can be seen that the Grüneisen parameter and the principal components of the Grüneisen tensor have similar temperature dependence. Another way of demonstrating that $\gamma(T)$ is strongly temperature dependent is to plot the volume coefficient of thermal expansion versus the molar heat capacity at constant pressure. Since the molar volume and the adiabatic compressibility are

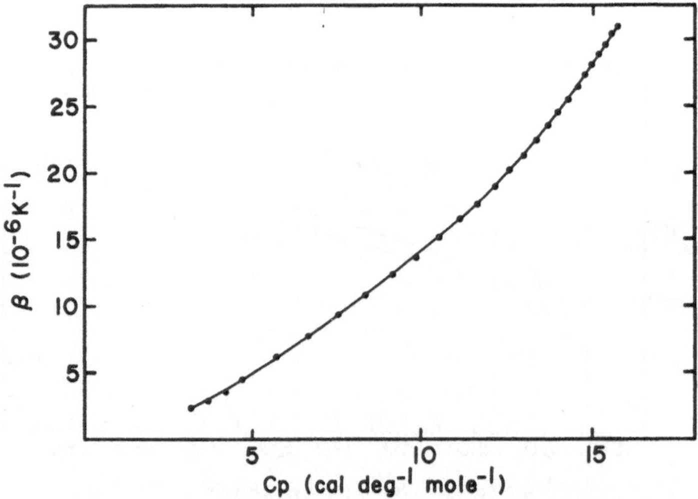

Fig. 4. The temperature dependence of the Grüneisen parameter for ZnF_2 as demonstrated by a plot of the β vs C_p.

only slightly temperature dependent, strong deviations from a straight line in the β versus C_p curve indicate a temperature dependence of γ(T). This has been presented in Figure 4.

ACKNOWLEDGMENTS

The authors would like to thank Dr. R. Fiegelson of Stanford University for graciously supplying the ZnF_2 samples used in this study.

REFERENCES

1. K.V.K. Rao and S.V.N. Naidu, Proc. Indian Acad. Sci. 58, 296 (1963).
2. J.W. Stout and S.A. Reed, J. Am. Chem. Soc. 76, 5279 (1954).
3. K.V.K. Rao, AIP Con. Proc. 17, 219 (1974).
4. J.S. Browder, J. Phys. Chem. Solids 36, 193 (1975).
5. K. Haefner, Ph.D. thesis (Univ. of Chicago, 1964).
6. D.S. Rimai, Phys. Rev. B 16, 4069 (1977).
7. J.W. Stout and E. Catalano, J. Chem. Phys. 23, 2013 (1955).

STUDIES ON LATTICE EXPANSION AND DEBYE TEMPERATURES OF MIXED CRYSTALS OF $Ca_xSr_{1-x}F_2$

S.V. Suryanarayana and G. Jayasagar

Physics Department, Osmania University
Hyderabad 500 007, India

INTRODUCTION

Crystals with the fluorite structure are probably structurally simplest except for those of alkali halides with the sodium chloride structure. A variety of investigations on the physical properties of alkaline earth halides crystallizing in the fluorite structure have been reported in the literature. Two- and three-component systems based on alkaline earth fluorides have been grown by melt techniques. An extensive review of various thermodynamic and defect properties of fluorite-type crystals is available (1). The system $Ca_xSr_{1-x}F_2$ forms a continuous series of solid solutions and obeys Vegard's law (2). A perusal of the literature indicates that there is no reported work on the thermal expansion, Debye-Waller factors, and Debye temperatures of the mixed crystals of alkaline earth fluorides. The present paper deals with such data on the CaF_2-SrF_2 mixed system, as a part of the general program of investigations on fluorite-based mixed crystals.

EXPERIMENTAL

Alkaline earth fluorides have high melting temperatures. These crystals can be grown by conventional melt techniques like the Czochralski and Bridgman-Stockbarger methods (3,4). In the present work, various compositions of the mixed system $Ca_xSr_{1-x}F_2$ were grown by the Bridgman-Stockbarger technique. The crystals were grown in a vacuum furnace specially fabricated for this purpose (5). The furnace operates at 10^{-5} torr vacuum and can reach temperatures up to 1800°C. Graphite heaters were used. The crucibles for the growth of single crystals were also made of graphite.

The samples necessary for the thermal expansion and intensity measurements were obtained by crushing single crystals, grown in the vacuum furnace, into fine powder of suitable particle size.

A back-reflection focusing camera (6) was used to collect the data on lattice parameters at various temperatures. The lattice parameters were evaluated using graphical extrapolation with an error function $\phi \tan \phi$. The coefficient of thermal expansion for each sample was evaluated by following a semianalytical approach suggested by Deshpande and Mudholkar (7).

The diffractometric traces required for the evaluation of the Debye-Waller factors and the Debye temperatures were collected on a PW 1730 X-ray generator fitted with a copper target and PW 1050/70 vertical goniometer. The raw intensity data were corrected for the usual geometrical and physical factors (8).

In a mixed crystal of the type $Ca_xSr_{1-x}F_2$, the metal atoms are distributed at random at the metal atom sites and thus are indistinguishable. Therefore, the probability of finding a Ca or Sr atom at any one of the metal atom positions is proportional to the atom fractions x and (1 - x), while the fluorine positions always are occupied by the fluorine atoms. Under the assumption of random distribution of the metal atoms at the cation sites, the atomic scattering factor of the metal ion (f_M) is evaluated using the following relation:

$$f_M = x\, f_{Ca} + (1 - x) f_{Sr}$$

This procedure of evaluating the scattering factors is similar to the method employed earlier for binary and ternary alloy systems (9-11).

The average B factor required to evaluate the Debye temperature is calculated from the individual B factors, obtained from the respective graphs of log (I_0/I_c) vs $\sin^2 \Theta/\lambda^2$. The details of the reduction of X-ray intensity data to evaluate the desired results are given by James (12) and Warren (13).

RESULTS AND DISCUSSION

From the data on the lattice parameters at various temperatures the coefficients of thermal expansion of CaF_2, SrF_2, and six intermediate compositions in the temperature range 30-405°C have been evaluated. Table 1 gives the results on thermal expansion, and the data are shown graphically in Fig. 1.

X-ray diffractometric data on CaF_2, SrF_2, and five intermediate compositions were collected and analyzed. The results are given in Table 2. This table lists the values of the lattice parameter at room temperature (a), the Debye-Waller factor for the metal ion (B_M), that for

Table 1. Data on Thermal Expansion of the CaF_2-SrF_2 System

Units: $\alpha \times 10^6$

Temp. °C	CaF_2	A	B	C	D	E	F	SrF_2
30	18.43	17.12	15.60	15.76	16.71	18.69	16.57	16.28
45	18.67	17.22	15.81	16.01	16.87	18.76	16.70	16.37
100	19.55	17.60	16.57	16.93	17.47	19.04	17.23	16.76
105	19.64	17.64	16.63	17.01	17.53	19.07	17.28	16.80
165	20.62	18.15	17.39	17.97	18.23	19.48	17.97	17.39
195	21.11	18.44	17.75	18.44	18.60	19.71	18.35	17.74
200	21.18	18.49	17.81	18.51	18.66	19.75	18.42	17.81
255	22.09	19.08	18.43	19.34	19.37	20.24	19.20	18.57
300	22.81	19.61	18.90	19.99	19.99	20.70	19.90	19.29
315	23.08	19.79	19.05	20.20	20.20	20.86	20.15	19.56
345	23.51	20.19	19.34	20.62	20.63	21.20	20.66	20.11
400	24.46	20.95	19.83	21.35	21.45	21.89	21.67	21.22
405	24.56	21.03	19.87	21.42	21.53	21.95	21.76	21.33

A = $Ca_{0.9}Sr_{0.1}F_2$ C = $Ca_{0.7}Sr_{0.3}F_2$ E = $Ca_{0.4}Sr_{0.6}F_2$

B = $Ca_{0.8}Sr_{0.2}F_2$ D = $Ca_{0.5}Sr_{0.5}F_2$ F = $Ca_{0.3}Sr_{0.7}F_2$

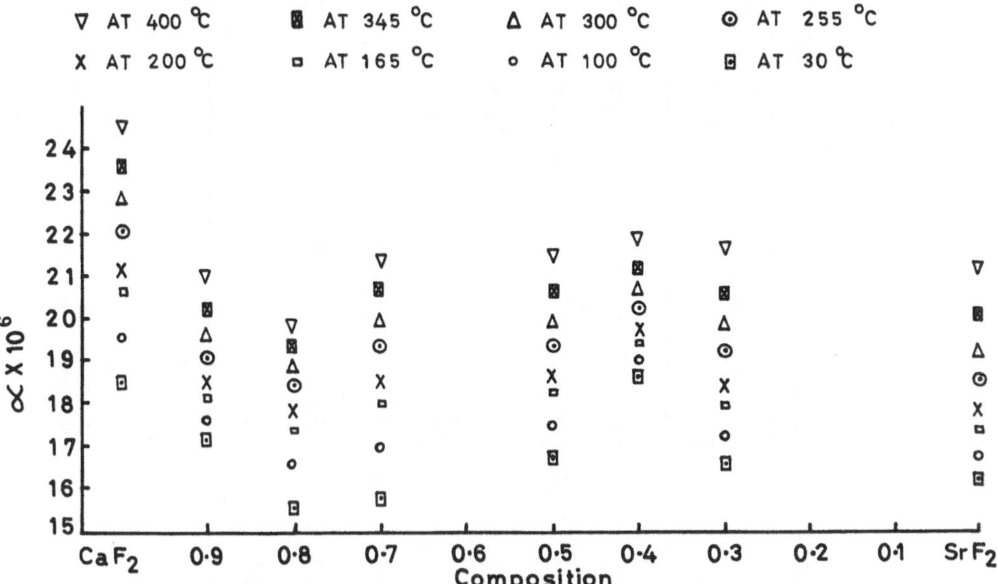

Fig. 1. Thermal expansion versus composition at various temperatures for the CaF_2-SrF_2 system.

Table 2. Summary of Results of the Diffractometric Work

Compound	a (Å)	B_M ($Å^2$)	B_F ($Å^2$)	$<\mu_M^2>^{1/2}$ (Å)	$<\mu_F^2>^{1/2}$ (Å)	Θ_M (°K)
CaF_2	5.463	0.750	0.700	0.169	0.163	438
$Ca_{0.9}Sr_{0.1}F_2$	5.494	0.733	0.860	0.167	0.181	399
$Ca_{0.7}Sr_{0.3}F_2$	5.569	0.717	1.267	0.165	0.219	356
$Ca_{0.5}Sr_{0.5}F_2$	5.637	0.567	1.400	0.147	0.231	346
$Ca_{0.4}Sr_{0.6}F_2$	5.675	0.600	1.267	0.151	0.220	345
$Ca_{0.1}Sr_{0.9}F_2$	5.765	0.666	1.317	0.159	0.224	317
SrF_2	5.795	0.733	1.230	0.167	0.216	309

the fluorine ion (B_F), the corresponding root-mean-square amplitudes of vibration ($<\mu_M^2>^{\frac{1}{2}}$ and $<\mu_F^2>^{\frac{1}{2}}$), and the Debye temperature Θ_M.

A study of Table 1 leads to the following observations about the thermal expansion behavior of the mixed system. For any particular composition, including the end members, the coefficient of expansion is found to increase with temperature. The rate at which α varies with temperature for various compositions is found to differ. Thus $Ca_{0.4}Sr_{0.6}F_2$ shows a minimum of 17.44% while $Ca_{0.7}Sr_{0.3}F_2$ shows a maximum variation of 35.9% in the temperature range covered.

The variation of α at any temperature between the two end members - CaF_2 and SrF_2 - is not very large. For any composition the temperature variation of α is nonlinear. While the general trend for any composition is to show a value of α between the limits of the end members, values less than those have also been observed. However, the differences are not large. The probable reason for this behavior may lie in the smaller differences in the melting points of the end members and also those of the intermediate compositions.

As the Ca content decreases, there is a corresponding decrease in α up to $Ca_{0.7}Sr_{0.3}F_2$, and later there is an increasing trend in the values of α, showing a maximum value for $Ca_{0.4}Sr_{0.6}F_2$. At this composition, the value of α observed is nearly equal to that observed for CaF_2. The mixed crystals beyond this composition (toward the SrF_2 end) appear to exhibit a tendency to lower values of α, approaching the value observed for SrF_2. Since the thermal expansion is an anharmonic property, it may not be normally expected that an additive law of expansion in the mixed system, similar to Vegard's law, applies also to a solid solution. There is a significant difference between the observed and the calculated values of α computed on the basis of an additive law.

Though it is not obvious why the particular composition $Ca_{0.4}Sr_{0.6}F_2$ shows a departure from the behavior of other compositions, as one goes from the CaF_2 to the SrF_2 side, a probable explanation may be the following.

It is generally known that vacancies also contribute to the coefficient of thermal expansion (14). In mixed crystals, in addition to the usual thermal vacancies, vacancies may also be produced because of the presence of a second type of atoms. As the concentration of the second component increases, one may expect more and more disorder in the host lattice compared to the end members. For a composition near a particular end member, the observed coefficient of expansion may be influenced by that particular member. This observation can be seen to be borne out by $Ca_{0.9}Sr_{0.1}F_2$ and $Ca_{0.3}Sr_{0.7}F_2$. However, the temperature variation of α at any particular composition, though non-linear, is difficult to explain on the basis of the influence of the end members alone. For crystals near the equimolar composition, the lattice may be much more disordered and the concentration of vacancies may be greater. Thus, the ionic contributions to α from lattice defects may also be significant.

A study of Table 2 leads to the following observations. The Debye-Waller factor for the cation (B_M) is found to decrease with decreasing content of CaF_2 in the mixed system until the equimolar composition is reached; thereafter the value shows an increasing trend and finally approaches the value of the other end member, i.e., SrF_2. On the other hand, the value of the Debye-Waller factor

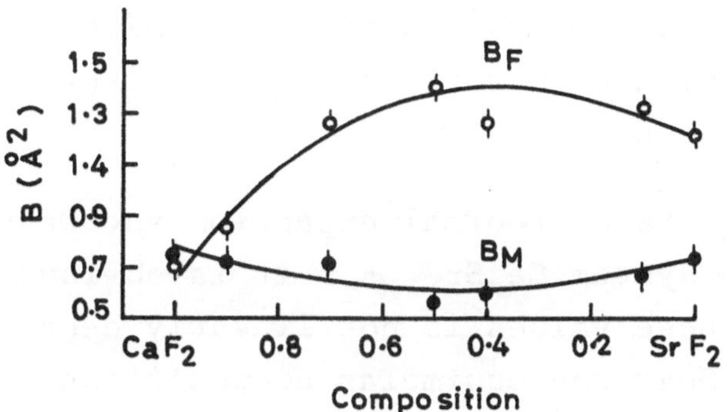

Fig. 2. Debye-Waller factor versus composition.

for fluorine (B_F) is found to increase with decreasing CaF_2 content, attains a maximum value of 1.4 Å^2 at about the equimolar composition, and then shows a decreasing trend, finally attaining the value of the Debye-Waller factor for fluorine in SrF_2, the other end member.

The variation of B factors with composition is shown graphically in Fig. 2. From the individual values of B factors, the average B factor was evaluated, and from this value the Debye characteristic temperature is calculated. It is found that Θ decreases with increasing SrF_2 content, as shown in Fig. 3. It is of interest to mention that Karlsson (15) observed a similar nonlinear and decreasing variation of Θ for the system KCl-KBr from his data on specific heats.

From the data on thermal expansion and Debye-Waller factors of the system $Ca_xSr_{1-x}F_2$, it is obvious that the variation of these values is not linearly dependent on composition. Near the equimolar compositions, the variations in the physical behavior of these crystals are interesting. The thermal expansion is an anharmonic property. As the temperature increases, certain Bragg intensities in the fluorite structure depart progressively from the predictions of the harmonic approximation (16-18). In addition, the density of defects may also differ as a function of composition. Thus, varying defect concentration values combined with varying thermal vibration amplitudes may give rise to a situation in which the observed coefficients of expansion are affected.

Thus it seems desirable to investigate the CaF_2-SrF_2 mixed system, in particular, to carry out a detailed examination of macroscopic and X-ray volume expansivity

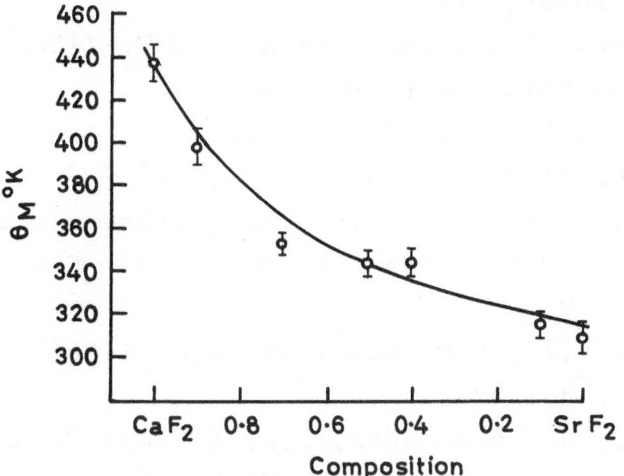

Fig. 3. Θ_M versus composition for the CaF_2-SrF_2 system.

measurements. Such data, combined with the data on Debye-Waller factors of these crystals at different temperatures, may help in understanding the thermal expansion behavior vis-à-vis the amplitudes of vibrations of atoms, and their contribution to the higher-order terms in the potential. Such work is currently in progress in our laboratory.

REFERENCES

1. Hayes, W. (1974), "Crystal with Fluorite Structure," Clarendon, Oxford.
2. Chernevskaya, E.G., and Anan'eva, G.V. (1966), Sov. Phys. Solid State 8, 169.
3. Stockbarger, D.C. (1949), J. Opt. Soc. Amer. 39, 731.
4. Rao, S.M.D. (1976), Ind. J. Phys. 50, 378.
5. Jayasagar, G. (1981), Ph.D. Thesis, Osmania University.
6. Sirdeshmukh, D.B. (1963), Ph.D. Thesis, Osmania University.
7. Deshpande, V.T., and Mudholkar, N.H. (1961), Ind. J. Phys. 35, 434.
8. Klug, H.P., and Alexander, L.E. (1974), "X-Ray Diffraction Procedures," John Wiley and Sons, New York.
9. Herbestein, F.H., Averbach, B.L., and Borie, B.N. Jr. (1956), Acta Cryst. 9, 466.
10. Herbestein, F.H., and Averbach, B.L. (1956), Acta Met. 4, 40.
11. Naidu, S.V.N., and Houska, C.R. (1971), J. Appl. Phys. 42, 4971.
12. James, R.W. (1967), "Optical Principles of the Diffraction of X-Rays," G. Bell and Son, London.
13. Warren, B.E. (1969), "X-Ray Diffraction," Addison-Wesley, Reading, Mass.
14. Simmons, R.O. (1968), Int. Symp. on Thermal Expansion of Solids, NBS, Washington.

15. Karlsson, Å.V. (1970), Phys. Rev. B2, 3332.
16. Willis, B.T.M. (1963), Proc. Roy. Soc. A 274, 122.
17. Willis, B.T.M. (1965), Acta Cryst. 18, 75.
18. Willis, B.T.M. (1969), Acta Cryst. A25, 277.

ANOMALOUS THERMAL EXPANSION OF SOME CHALCOPYRITE TYPE SEMICONDUCTOR COMPOUNDS

K. Satyanarayana Murthy, P. Kistaiah, Y.C. Venudhar, Leela Iyengar, and K.V. Krishna Rao+

Department of Physics
Osmania University
Hyderabad - 500 007, India

INTRODUCTION

It is known that ternary compounds $A^{I}B^{III}C_2^{VI}$ are semiconductors, possess the chalcopyrite structure and belong to the space group D_{2d}^{12} -$I\bar{4}2d$[1,2]. Recently these compounds have received considerable attention because of their possible applications in electro-optical devices[3,4]. However, only a little work has been reported on the thermal properties of these semiconductors. Recent measurements of the thermal expansion coefficients of some of these compounds [5-7] have shown that the linear thermal expansion coefficients α_a and α_c of the lattice parameters a and c, respectively, are anisotropic. In some cases even negative values for α_c have been found [5,6]. Besides, in this tetragonal chalcopyrite structure (Figure 1) the two kinds of cations make an ordered sublattice which may result in a tetragonal distortion of the lattice defined as $\partial = 2-c/a$, c and a being the lattice parameters. It is interesting to note that the anisotropy of the thermal expansion coefficients results in a temperature dependence of the tetragonal distortion δ which seems to depend in a simple manner on the distortion δ itself[8].

The knowledge of the thermal expansion coefficients, of its anisotropy, and of the temperature coefficients of the tetragonal distortion is quite important in single crystal growth experiments [5], in the choice of substrate materials suitable for epitaxial growth of these compounds[9], and in the understanding of the temperature

+ Present address : Department de Physique
University de Constantine, Constantine, Algeria

dependence of the electronic properties[7]. In view of this technological as well as physical interest in the temperature dependence of the lattice parameters on the one hand and the lack of corresponding experimental data for many of these compounds on the other hand, it was thought desirable to undertake the systematic thermal expansion study on $A^{I}B^{III}C_2^{VI}$ ternary compounds. This paper gives an account of the results obtained on four members of this family of compounds, namely, $AgInSe_2$, $AgGaTe_2$, $AgGaSe_2$ and $AgGaS_2$.

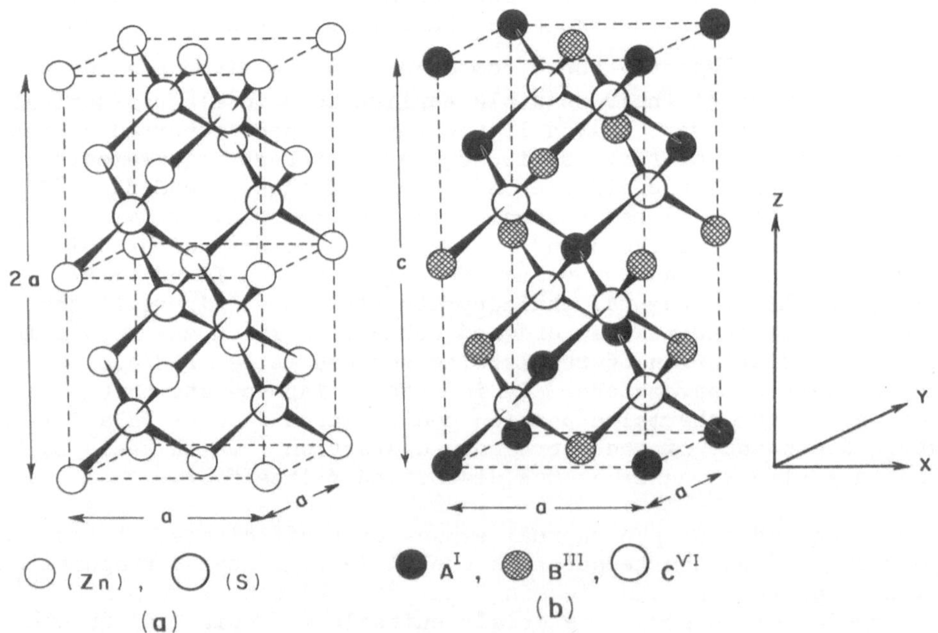

Figure 1 : The chalcopyrite type structure (b) compared to two unit cells of the zinc blende structure (a)

EXPERIMENTAL

X-ray powder diffraction patterns of each substance at temperatures ranging from 28°C to about 480°C were obtained with a Unicam 19 cm high temperature powder camera and CuK_α radiation. Temperature control to within 2°C was obtained with the use of a voltage stabilizer and variac. Silver was used as a standard substance to estimate the specimen temperature.

Every attempt was made to record high angle reflections for the precision determination of the lattice parameters. In each case, about ten diffraction lines with Bragg angles ranging from 60 to 80° were used to evaluate the lattice parameters applying Cohen's(10) analytical method. The details of the experimental arrangement and the evaluation of the precision lattice parameters and the coefficients of thermal expansion were given in earlier publications(11,12) from this laboratory.

RESULTS

The preliminary results obtained in individual crystals are presented in recent publications(13-15) from this laboratory. Table-I gives the data on the lattice parameters, the principal coefficients of thermal expansion, the average linear thermal expansion coefficients, the coefficients of volume expansion, the axial ratios and the anisotropy coefficients for $AgInSe_2$, $AgGaTe_2$, $AgGaSe_2$ and $AgGaS_2$ at the lowest and at the highest temperatures of the range covered. The variation of the lattice parameters with temperature in these compounds is shown in Figures 2, 3 and 4. In evaluating the lattice parameters, independent measurements and calculations have been made and the deviations of the individual values from the mean have been found to be about ± 0.0001 Å in the a parameter and about ± 0.0002 to ± 0.0003 Å in the c parameter.

In each case, using the least squares method, the coefficients of thermal expansion α_a and α_c of the lattice parameters a and c respectively, at different temperatures have been represented by equations of the type

$$\alpha_T = A + BT + CT^2 + DT^3 \tag{1}$$

where A,B,C and D are constants and T is the temperature in degrees Celsius. Table-II gives a compilation of the thermal expansion data obtained in the present study. The accuracy in the values of the thermal expansion coefficients as estimated from an error equation(16) has been found to be about three per cent.

TABLE I
Some data on $AgInSe_2$, $AgGaTe_2$, $AgGaSe_2$ and $AgGaS_2$ ternary chalcopyrite type compounds

Compound	Tempe-rature (°C)	a(Å)	c(Å)	$10^6\alpha_a$	$10^6\alpha_c$	$10^6\alpha$	$10^6\beta$	c/a	α_c/α_a
$AgInSe_2$	28	6.1008	11.6908	-3.11	4.73	-0.50	-1.49	1.9163	-1.52
	478	6.0907	11.6999	-4.08	-2.96	-3.71	-11.12	1.9210	0.73
$AgGaTe_2$	28	6.3197	11.9834	15.10	-2.83	9.12	27.37	1.8962	-1.87
	394	6.3567	11.9548	16.45	-14.86	6.01	18.04	1.8807	-0.90
$AgGaSe_2$	28	5.9912	10.8851	16.40	-13.30	6.50	19.50	1.8169	-0.81
	478	6.0399	10.8066	21.38	-16.94	8.61	25.82	1.7892	-0.79
$AgGaS_2$	28	5.7556	10.2989	8.88	-7.76	3.33	10.00	1.7894	-0.87
	394	5.7785	10.2584	14.60	-18.43	3.59	10.77	1.7753	-1.26

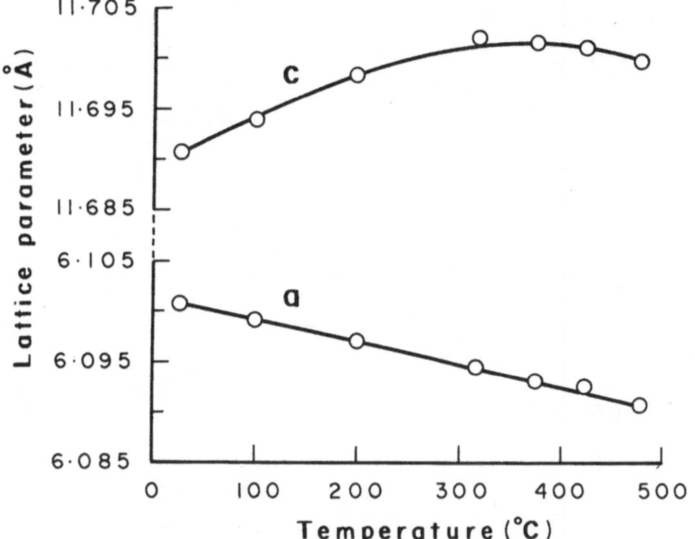

Figure 2 : Variation of lattice parameters of $AgInSe_2$ with temperature

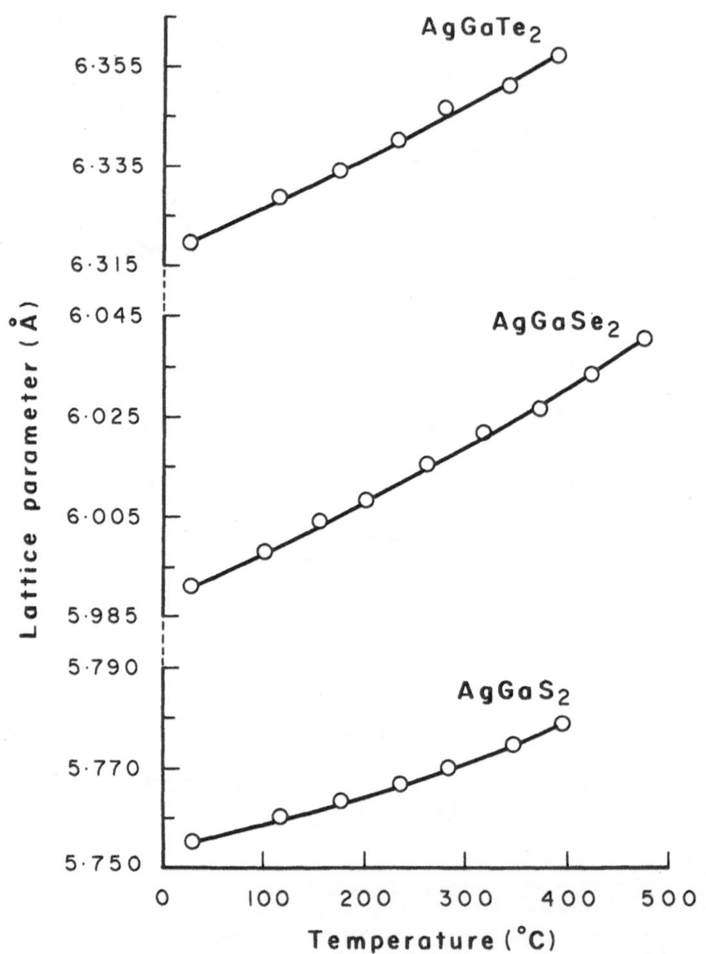

Figure 3 : Variation of a parameter with temperature

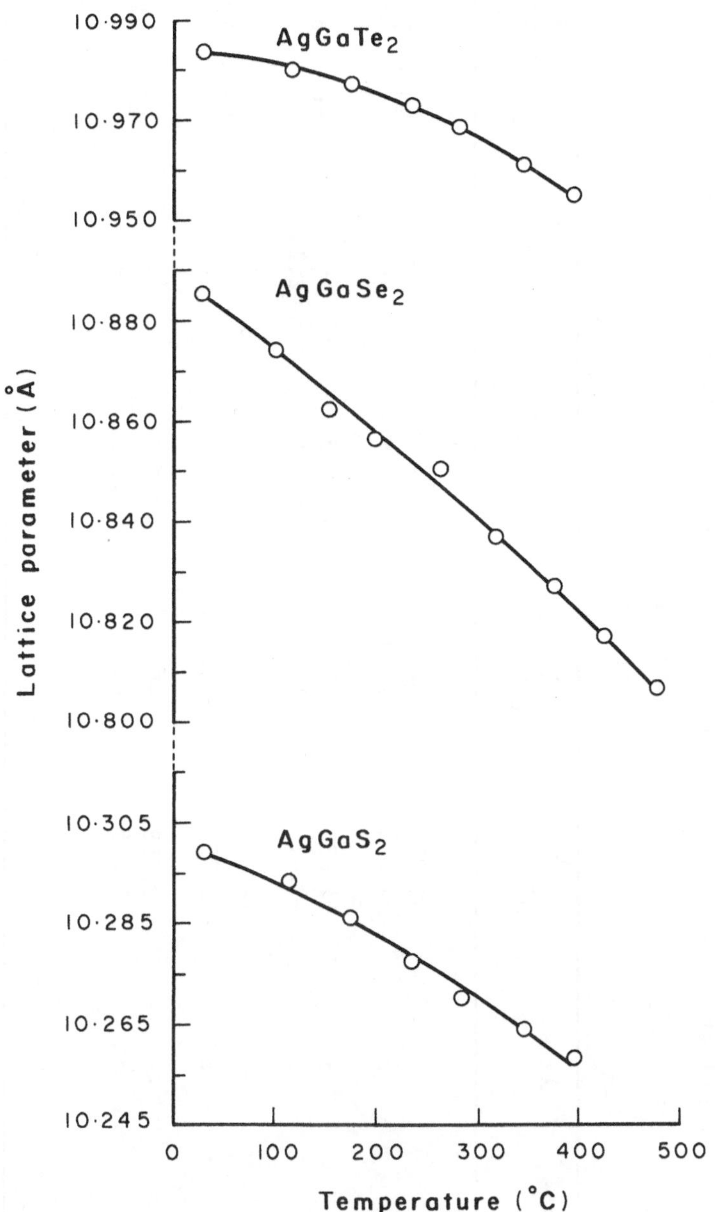

Figure 4 : Variation of c parameter with temperature

TABLE II

Compilation of thermal expansion data on $AgInSe_2$, $AgGaTe_2$, $AgGaSe_2$ and $AgGaS_2$ ternary compounds

$$\alpha_T = A + BT + CT^2 + DT^3$$

Compounds	Axis	A $(10^{-6}/^oC)$	B $(10^{-9}/^oC^2)$	C $(10^{-11}/^oC^3)$	D $(10^{-13}/^oC^4)$	Range of Temperature (^oC)
$AgInSe_2$	a	-3.052	-2.142	-	-	28 - 478
	c	4.842	-3.364	-2.709	-	
$AgGaTe_2$	a	14.889	7.854	-0.985	-	28 - 394
	c	-2.414	-15.962	5.070	-2.294	
$AgGaSe_2$	a	16.427	-1.708	2.523	-	28 - 478
	c	-12.729	-20.963	2.543	-	
$AgGaS_2$	a	8.858	-0.354	3.790	-	28 - 394
	c	-7.665	-1.512	-6.551	-	

DISCUSSION

It can be seen from Table-I that although all the crystals belong to the same structure type, their thermal behaviour shows large variations both in the values of the coefficients of thermal expansion and in their anisotropy. However, except $AgInSe_2$, the other three compounds, $AgGaTe_2$, $AgGaSe_2$ and $AgGaS_2$, show a similar thermal behaviour. In these crystals, the coefficient of expansion along the c-axis (α_c) is negative and increases numerically with increasing temperature. In all the three cases, the coefficient of volume expansion is positive but increases with temperature in the case of $AgGaSe_2$ and $AgGaS_2$ while it decreases in $AgGaTe_2$. It can be also seen from Table-I that for these compounds the c/a value decreases with increasing temperature giving rise to an increase in the tetragonal distortion of the lattice with temperature. In what follows, the thermal behaviour of these three compounds is explained on the basis of their structure.

In the tetragonal chalcopyrite structure as shown in Figure 1, each cation is approximately tetrahedrally coordinated with four anions while each anion is tetrahedrally coordinated with two cations of either type. From Figure 1 it is apparent that the structure is very closely related to that of zinc-blende. However, the presence of two different cations, from different columns of the periodic table, gives rise to two cation-anion bonds, I-VI and III-VI. The III-VI bond is more covalent(8) and hence is stiffer than the I-VI bond, leading to a displacement of the anion relative to its equilibrium position from zinc-blende. As a consequence of this displacement of anion, a tetragonal distortion defined earlier results in this structure. The increase in the tetragonal distortion in these compounds (Table-I) with temperature can be interpreted in terms of the thermal expansion of the individual I-VI and III-VI bonds by applying the Abrahams and Bernstein (17) relations to the present results. Work by these investigators produced the presently accepted picture of the $A^{I}B^{III}C_2^{VI}$ compounds as consisting of an almost perfect BC_4 tetrahedral arrangement with tetrahedral angles within 2% of the ideal value of 109.47°. Assuming an almost perfect BC_4 tetrahedron, the anion free parameter 'x' and the bond lengths can be expressed in terms of the lattice parameters, a and c,

$$x = 0.5 - \left[\frac{c^2}{32a^2} - \frac{1}{16} \right]^{\frac{1}{2}} \tag{2}$$

$$A - C = \left[a^2x^2 + \frac{4a^2 + c^2}{64} \right]^{\frac{1}{2}} \tag{3}$$

$$B - C = \left[a^2(\tfrac{1}{2} - x)^2 + \frac{4a^2 + c^2}{64} \right]^{\frac{1}{2}} \tag{4}$$

$$c = \frac{8}{\sqrt{3}} (B-C) \tag{5}$$

Here the 'x' parameter denotes the location of the group VI anion relative to the group I and III cations by the atomic distribution as given by Shay and Wernick[3] in the chalcopyrite unit cell. The calculated bond expansion coefficients using the above equations are listed in Table-III along with the temperature coefficients of the 'x' parameter and the axial ratio of the compounds studied in the present investigation. The thermal expansion of the bonds given in Table-III for $AgGaTe_2$, $AgGaSe_2$ and $AgGaS_2$ indicate the reason for the increase in the tetragonal distortion with temperature. The more ionic I-VI bonds have higher thermal expansion coefficients than the covalent III-VI bonds, so that, the group VI atoms move relatively nearer the group III cations on heating, thus increasing the tetragonal distortion through equation(2). As pointed out by Abrahams and Bernstein[17], the AC_4 tetrahedron is more oriented along the yz-plane while the BC_4 tetrahedron is more oriented along the xy-plane of the chalcopyrite lattice. Because of their bond nature as mentioned earlier, the thermal expansion along the a-direction (α_a) is greater than that along the c-direction in these compounds.

As already mentioned, one of the coefficients of expansion α_c is negative for $AgGaTe_2$, $AgGaSe_2$ and $AgGaS_2$. A survey of the literature[18-20] shows that many crystals exhibit negative coefficients of expansion in the low temperature region and the negative coefficients in these cases have been explained as due to the preponderance at low temperatures of acoustic vibrations satisfying the condition $\partial V/\partial p < 0$. But, in the case of $AgGaTe_2$, $AgGaSe_2$ and $AgGaS_2$, the negative coefficients of expansion are in the high temperature region and the same explanation may not hold good. The negative coefficient of expansion (α_a) in the case of layer type compounds like calcite[21] and graphite[22] in the high temperature region has been interpreted in terms of a Poisson contraction in that direction arising from the large coefficient of expansion in a perpendicular direction (α_c). But this explanation does not hold good in the case of the present compounds which are not of layer type.

The tetrahedral coordination in these materials implies that the bonding is primarily covalent with sp^3 hybrid bonds prevalent, although there is some ionic character present because of the two different types of cations. Recent band structure calculations[3] of these compounds showed that the d-electronic states of the A^I cations play a dominant role on their valence band structure and on the formation of hybrid bonds. The d-electrons in these crystals may give rise to a lattice potential having an unusual shape along the c-direction which results in the anomalous thermal behaviour.

TABLE III

The mean values of temperature coefficients of the tetragonal distortion, the 'x' parameter and the bond expansion coefficients of $AgInSe_2$, $AgGaTe_2$, $AgGaSe_2$ and $AgGaS_2$ ternary compounds. The suffix 'o' denotes the room temperature values

Parameter ($10^{-6}/^{o}C$)	$AgInSe_2$ +	$AgGaTe_2$ ++	$AgGaSe_2$ +	$AgGaS_2$ ++
$\frac{1}{(c/a)_o} \frac{\Delta(c/a)}{\Delta T}$	5.42	- 22.33	- 33.76	- 21.53
$\frac{1}{x_o} \frac{\Delta x}{\Delta T}$	- 10.01	41.47	58.83	36.79
$\bar{\alpha}_{I-VI}$	- 5.87	25.93	35.72	22.60
$\bar{\alpha}_{III-VI}$	1.73	- 6.74	- 16.03	- 10.84

\+ - Range of Temperature : 28 to 478°C

++ - Range of Temperature : 28 to 394°C

Further, the observed anomalous thermal behaviour of $AgGaTe_2$, $AgGaSe_2$ and $AgGaS_2$ is in reasonable accordance with the findings of Neumann[23], who has derived some empirical relations for the thermal expansion coefficients of these compounds using a simple model based on the peculiarities in their chalcopyrite structure. It is also evident from these three compounds that the thermal expansion anisotropy, α_c/α_a, is a continuous function of c/a, increasing with decreasing lattice constant ratio c/a, and that α_c becomes negative below a critical c/a value, approximately equal to 1.90 as pointed out by Neumann[23].

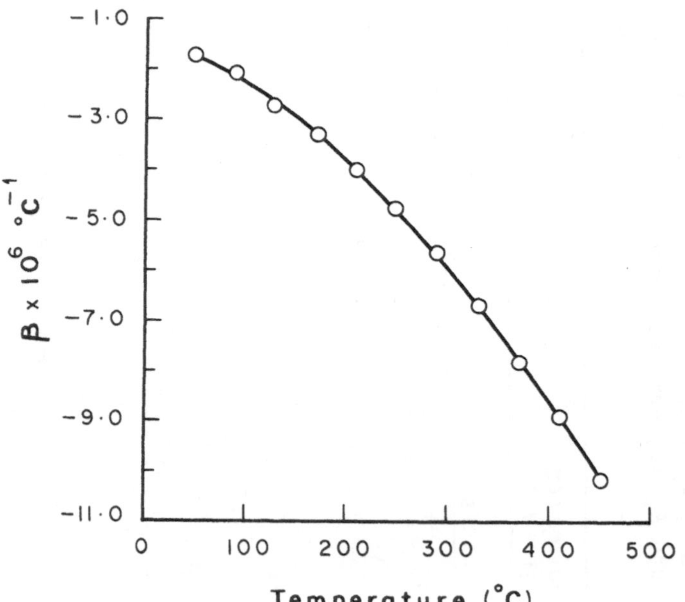

Figure 5 : Variation of coefficient of volume expansion of $AgInSe_2$ with temperature

Although the compound $AgInSe_2$ crystallizes in the same structure, its thermal behaviour is unlike that of the rest of the compounds studied in the present investigation. For this compound the coefficient of expansion along the a-axis is negative and its numerical value increases with temperature. Figure 5 shows the variation of the coefficient of volume expansion (β) of $AgInSe_2$ with temperature. It can be seen that the coefficient of volume expansion increases negatively with increasing temperature. Though many crystals exhibit a negative thermal expansion along one axis, their coefficients of volume expansion are usually positive especially at high temperatures. Some crystals having diamond structure like silicon and germanium show negative volume expansion at low temperatures[18,19]. So far only three compounds, β-AgI, thorium fluoride and mercurous chloride were reported to have negative volume expansion at high temperatures[24-26]. Thus $AgInSe_2$ is unusual in showing at high temperatures a negative volume expansion. At present no convincing explanation can be found for the very unusual thermal behaviour of this compound. It is felt that very much more accurate experimentation, particularly on the structure of this compound at different temperatures and on the electrical properties of this material, would be necessary before any convincing explanation could be offered for its anomalous thermal behaviour. An attempt to explain this unusual behaviour of $AgInSe_2$ along these directions is under way.

ACKNOWLEDGEMENTS

The authors wish to thank Dr. B.Tell of the Bell Laboratories, U.S.A., for providing the crystals used in the present study. They are also thankful to the University Grants Commission and Department of Science and Technology, New Delhi, India for their financial support. One of the authors (P.Kistaiah) is grateful to the Council of Scientific and Industrial Research, New Delhi, India for the award of a Senior Research Fellowship.

REFERENCES

1. Hahn, H., Frank, G., Klingler, W., Mayer, A.D. and Storger, G., (1953), Z. Anorg. Allgem. Chem. 271, 153.
2. Sandrock, R. and Treusch, J., (1964), Z. Naturforsch. A 19, 844.
3. Shay, J.L. and Wernick, J.H., (1975), 'Ternary Chalcopyrite Semiconductors : Growth, Electronic Properties and Applications', Pergamon Press, New York.
4. Thwaites, M.J., Tomlinson, R.D. and Hampshire, M.J., (1977), Solid State Commun. 23, 905.
5. Iseler, G.W., (1977), J.Cryst. Growth, 41, 146

6. Korczak, P. and Staff, C.B., (1974), J. Cryst. Growth, 24/25, 386.
7. Yamamoto, N., Horinaka, H. and Miyauchi, T., (1979), Jap. J. Appl. Phys. 18 , 255.
8. Weaire, D. and Noolandi, J., (1975), J. Phys. Suppl. 36, C3 - 27.
9. Bachmann, K.J., Beuhler, E., Shay, J.L. and Wagner, S., (1975), Z. Phys. Chem. NF 98 , 365.
10. Cohen, M.U., (1935), Rev. Sci. Instrum. 6 , 68.
11. Krishna Rao, K.V., Nagender Naidu, S.V. and Leela Iyengar, (1973), J. Appl. Cryst. 6 , 136.
12. Krishna Rao, K.V., (1969), 'Physics of the Solid State', Academic Press, New York, pp. 415.
13. Kistaiah, P., Venudhar, Y.C., Murthy, K.S., Iyengar, L. and Rao, K.V.K., (1981), J. Less-Common Met. 77, p 9.
14. Kistaiah, P., Venudhar, Y.C., Murthy, K.S., Iyengar, L. and Rao, K.V.K., (1981), J. Less-Common Met. 77, p 17.
15. Kistaiah, P., Venudhar, Y.C., Murthy, K.S., Iyengar, L. and Rao, K.V.K., (1981), J. Mat. Sci. (In Press).
16. Deming, W.E., (1943), 'Statistical Adjustment of Data', Wiley, New York, pp. 38.
17. Abrahams, S.C. and Bernstein, J.L., (1973), J. Chem. Phys. 59 , 5415.
18. Smith, T.F. and White, G.K., (1975), J. Phys. C : Solid State Phys. 8, 2031.
19. Novikova, S.I., (1966), 'Semiconductors and Semimetals', 2 , 41.
20. Soma, T., (1977), J. Phys. Soc. Japan, 42 , 1491.
21. Krishna Rao, K.V., Nagender Naidu, S.V. and Murthy, K.S., (1968), J.Phys. Chem. Solids, 29, 245.
22. Nelson, J.B. and Riley, D.P., (1945), Proc. Phys. Soc. 57, 477.
23. Neumann, H., (1980), Kristall and Technick, 15 , 849.
24. Lawn, B.R., (1964), Acta Cryst. 17 , 1341.
25. Barns, R.L., (1977), Mat. Res. Bull. 12 , 327.
26. Venudhar, Y.C., Prasad, T.R., Murthy, K.S., Iyengar, L. and Rao, K.V.K., (1981), Thermal Expansion, 7 (In Press).

LATTICE THERMAL EXPANSION OF LANTHANUM OXYCHLORIDE

T. Pallayya and V.T. Deshpande

Department of Physics, Osmania University
Hyderabad 500 007, India

INTRODUCTION

Lanthanum oxychloride and many of the oxychlorides of rare earth elements, crystallize in the space group p/4nmm of the tetragonal system (1). All these crystals have a PbFCI-type layer structure. Each complex layer consists of a central sheet of oxygen atoms, arranged on a body centered square lattice perpendicular to the c-axis of the tetragonal crystal, (Fig. 1). On either side of this sheet there are other sheets, first of the metal atoms and the next of chlorine atoms, both arranged on primitive square lattices (2). Two adjacent complex layers have their chlorine atom sheets facing each other. This peculiarity of the structure involves interesting problems about the interatomic bondings inside a layer and in between layers. Some information on the nature and magnitudes of these binding forces is likely to be obtained by a study of the directional thermal expansion of these crystals. Again, the cationic radii in these rare earth compounds and hence the unit cell volumes of these crystals show a regular decrease with increasing atomic number of the metal atoms (3). There have been no reported studies on the effect of rise of temperature on the lattice parameters and hence the unit cell volumes of these crystals. In view of this, a plan has been drawn to study systematically the temperature variation of the lattice parameters and the directional coefficients of thermal expansion of the crystals of some rare earth oxychlorides. The present paper reports the work done on the first member of this group.

EXPERIMENTAL

LaOCl was prepared by heating an aqueous solution of $LaCl_3$ to moderately high temperature (4). 99.99% pure $LaCl_3$, obtained from Indian Rare Earths Limited, was dissolved in double distilled water and the solution was gradually heated to dryness and then to 500°C in an oven for 48 hours and then to 1000°C for another period of 48 hours. It has been reported by Wendlandt (5) that hydrated $LaCl_3$ is completely hydrolyzed to LaOCl above 680°C, and that the salt does not decompose up to 850°C. It was observed in the present study that after annealing the salt at 1000°C good quality X-ray powder pictures were obtained, which could be completely indexed as those of LaOCl, and there was no indication of any decomposition.

A symmetrical focusing back-reflection camera was used for the high temperature X-ray diffraction work. The design and working of this camera have been described earlier by Sirdeshmukh and Deshpande (6). It could record pictures up to 600°C. The temperatures could be measured with an accuracy of ±1°C with the help of a calibrated thermo-couple galvanometer combination. Powder photographs were taken at eight different points between room temperature and 520°C using CuK_α radiation. Unambiguous indexing of a large number of high angle lines was achieved by an iterative procedure. Starting from relatively low angle lines of a Debye-Scherrer picture which could be clearly indexed and using analytical extrapolation techniques to obtain lattice parameter values, more and more high angle lines could be unambiguously indexed in successive stages. The powder lines used to evaluate the final values of the lattice parameters were $(118)_{\alpha_1\alpha_2}(501)_{\alpha_1\alpha_2}(510)_{\alpha_1\alpha_2}(502)_{\alpha_1\alpha_2}(208)_{\alpha_1\alpha_2}$ and $(512)_{\alpha_1\alpha_2}$. All these lines have their Bragg angles above 69°. Cohen's (7) least-squares technique was used along with $\phi \tan \phi$ as the error factor, where ϕ is $(180 - 2\theta)$. Standard errors in the lattice parameters were calculated by the method of Jette and Foote (8). The maximum error in both the parameters observed was 0.001 Å, which was taken to be the standard error in both the parameters at all the temperatures.

RESULTS

The values of the lattice parameters at different temperatures are given in Table I. Temperature-parameter graphs indicated that the variation of both the parameters with temperature was linear within the limits of experimental errors. Least-squares fitting of straight lines to these data gave the following expressions for the temperature dependence of the two parameters:

Table I. Value of the Lattice parameters of Lanthanum Oxychloride at different temperatures

Temperature (^oK)	a(Å)	c(Å)
301	4.1209	6.8795
375	4.1248	6.8859
488	4.1313	6.8959
576	4.1354	6.9028
640	4.1411	6.9111
691	4.1429	6.9132
735	4.1457	6.9184
793	4.1474	6.9226

$$a_t = 4.1193 + 5.60 \times 10^{-5}\ t$$

$$c_t = 6.8769 + 8.87 \times 10^{-5}\ t$$

Here, a_t and c_t are the values of the lattice parameters in Å at t^oC.

The principal coefficients of thermal expansion over the range of temperature covered in the study were therefore taken to be independent of temperature. Their values are

$$\alpha_a = 13.60 \times 10^{-6} / {}^oC$$

$$\alpha_c = 12.90 \times 10^{-6} / {}^oC$$

DISCUSSION

Table II shows a comparison of the room temperature values of the lattice parameters obtained in the present work with the data from literature. It is found that the results of the present study are in good agreement with earlier results.

The results on the thermal expansion brings out two interesting aspects of the behaviour of LaOCl. The first is that both the principal coefficients are independent of temperature over a fairly wide range. The second point is that the values of the two coefficients are almost equal. These results can be interpreted to suggest broadly that the crystal is very stable and remains firmly bound over the whole range of temperature and also that the binding forces along and perpendicular to the complex layers are not much different from each other. Binding of atoms in any complex layer of the structure is provided by bonds between central oxygen atoms and the metal atoms on either side, which in turn are bonded to the chlorine atoms of the outer sheets. Each oxygen atom is bonded to

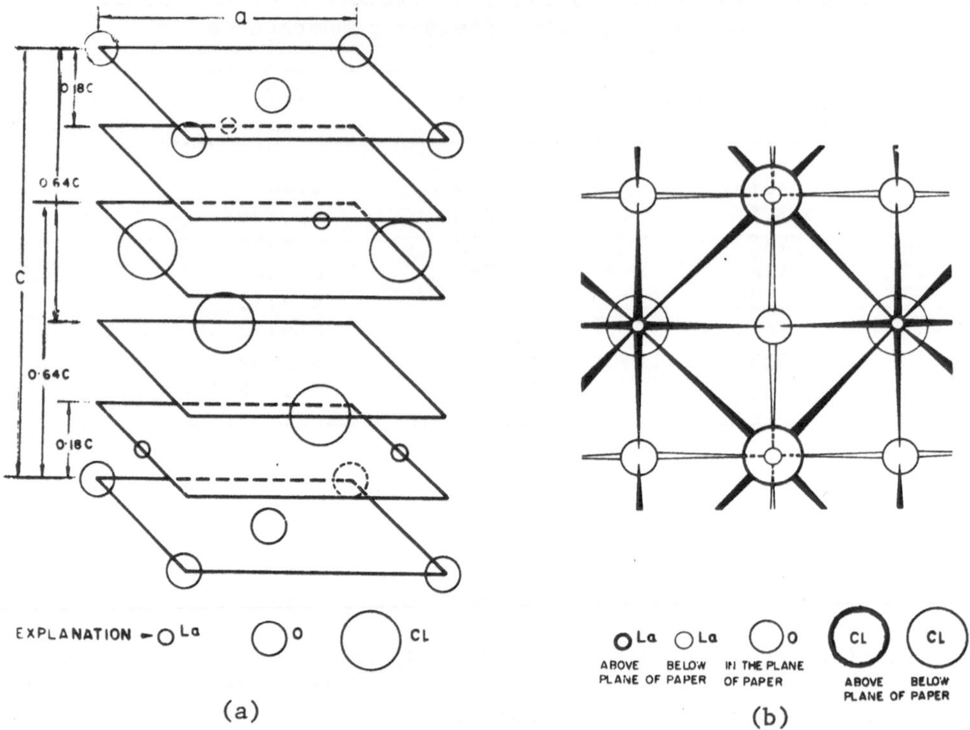

Fig. 1. (a) Unit Cell of LaOCl showing sequence of sheets; (b) Atomic arrangement in a complex layer of the LaOCl structure.

Table II. Comparison of the lattice parameters of Lanthanum Oxychloride at room temperature

Source	a(Å)	c(Å)
National Bureau of Standards (9)	4.120	6.882
Templeton et al.(3)	4.119 ± 0.002	6.883 ± 0.004
L.G.Sillen et al.(1)	4.117	6.879
S.Fried (10)	4.121 ± 0.003	6.885 ± 0.009
Present Study	4.1209 ± 0.001	6.8795 ± 0.001

two metal atoms from the sheet below it and two other metal atoms above it. Each metal atom has four oxygen neighbours on one side and four chlorine neighbours on the other (Fig.1) (2). The complex layers thus appear to be quite firmly bound by these linkages. However, the binding mechanism between layers does not seem to be so clear. The sheet of chlorine atoms from one layer is face to face with the sheet of chlorine atoms from the adjacent layer. The chlorine-chlorine distance in a layer is known to be larger than the Cl-Cl distance between layers. The former is 4.11 Å, while the latter is only 3.46 Å (2). Further, the second of these values is slightly smaller than twice the value of Van der Waals radius of chlorine, which is known to be 1.8 Å (11). These considerations may point out to the possibility of forces of covalent nature existing between chlorine atoms of adjacent layers. Another aspect of the structure is that the metal-chlorine distances from the same layer are 3.18 Å, while there is one chlorine atom from an adjacent layer just above or below a metal atom at a distance of 3.14 Å. This suggests the existence of an interlayer La-Cl bonding which may also be contributing to hold the layers together (2). A detailed quantitative analysis of these forces seems to be difficult at the moment. However, the results of thermal expansion measurements can be considered to indicate the near equality of the interlayer and intralayer binding forces.

REFERENCES

1. L.G.Sillen and A.L.Nylander, The Crystal Structure of LaOCl, LaOBr and LaOI, Svensk, Kemi. Tidskr., 53:367 (1941)
2. A.F.Well, "Structural Inorganic Chemistry", Third Edition, Oxford University Press, 391. (1962)
3. D.H.Templeton and Carol H.Dauben, Crystal Structures of Rare Earth Oxychlorides, J.Amer. Chem. Soc., 75:6069 (1953)
4. J.W.Mellor, "A Comprehansive Treatise on Inorganic and Theoretical Chemistry", Longmans, Green, London, 5:641 (1924)
5. Wesley W.Wendlandt, The Thermal Decomposition of Yttrium, Scandium, and some Rare-Earth Chloride Hydrates, J.Inorg. Nucl. Chem., 5:118 (1957)
6. D.B.Sirdeshmukh, and V.T.Deshpande, Design of A Symmetric Focusing Camera for Accurate Lattice Parameter Determination, Proc. Ind. Natl. Sci. Acad. 38:167 (1972)
7. M.U.Cohen, Precision Lattice Constants from X-Ray Powder Photographs, Rev. Sci. Instruments 6:68 (1935)
8. E.R.Jette, and F.Foote, Precision Determination of Lattice Constants, J.Chem Phys., 3:605 (1935)
9. H.E.Swanson et al., Standard X-ray Diffraction Powder Patterns, National Bureau of Standards, Circular., 7:22 (1958)
10. S.Fried, W.Hagemann and W.H.Zachariasen, J.Amer.Chem.Soc 75:771 (1950)
11. L.Pauling, "The Nature of Chemical Bond" Oxford University Press 189 (1952)

THERMAL EXPANSION BEHAVIOR OF METAL MATRIX COMPOSITES

B. K. Min and F. W. Crossman

Lockheed Palo Alto Research Laboratory
3251 Hanover Street, Palo Alto, CA 94304

INTRODUCTION

In recent years, graphite fiber reinforced metal matrix composites have received attention because these materials are potentially ideal to meet the need of dimensionally stable structures and optical systems. These materials have high specific strength and modulus, good thermal and electrical conductivity, low moisture absorption and dimensional stability under long term thermal exposure.

These composites exhibit low thermal expansions in the fiber direction due to near zero thermal expansion characteristics of graphite fibers. The large thermal expansion mismatch between the graphite fibers and the metallic matrices such as aluminum and magnesium generate thermal stresses in each constituent large enough even under moderate temperature ranges to produce plastic flow and creep of matrix materials, which may then alter and/or degrade their subsequent thermal and mechanical properties. Considerable research has been conducted in this area to predict and to measure the thermal stresses and to characterize the thermal expansion characteristics.

In this paper, we report the results of thermal expansion measurements for unidirectional graphite-aluminum and graphite-magnesium. An analysis that computes the internal stresses and the thermal expansion properties of unidirectional metal matrix composites is first presented. The experimental results are then shown together with the theoretical predictions to emphasize the effects considered in the analysis.

THERMAL EXPANSION ANALYSIS

During temperature change, the thermal expansion mismatch between the matrix and fiber may generate large thermal stresses in the matrix and fiber. For a well bonded unidirectional composite with a ductile metal matrix, the thermal stress may induce elastic, plastic, and creep deformation of the matrix. This deformation of the matrix must be accommodated by the deformation of the fiber. This results in the characteristic nonlinear, path-dependent thermal expansion behavior of the metal matrix composites. Such effects have been discussed for various fiber-metal matrix systems[1-5].

To rigorously analyze the thermal deformation of unidirectional composites may require a micromechanistic model which describes the detailed inhomogeneous deformation of the matrix around the fiber during thermal change[6-9]. However, the results of such analyses depend on an assumed geometric arrangement of the matrix and fiber[10-12]. A more simplistic engineering approach has been to use 'mechanics of materials' or energy approach to analyze elastic-plastic thermal expansion[13-16]. These were shown to provide reasonably accurate predictions for longitudinal and transverse thermal expansion. Uses of similarly simple continuum models for the analysis of elastic-plastic deformation of unidirectional composites under axisymmetric and plane stress assumptions were recently advanced[17,18]. In this section we analyze the thermal expansion of unidirectional metal matrix composites using a plane stress continuum model. The coordinates x and y are in the direction parallel and perpendicular to the fiber respectively. $d\underset{\sim}{\sigma} = [d\sigma_x, d\sigma_y, d\tau_{xy}]^T$, $d\underset{\sim}{\varepsilon} = [d\varepsilon_x, d\varepsilon_y, d\gamma_{xy}]^T$ denote in-plane stress and strain components of the composite. With superscripts m and f these symbols also denote the corresponding quantities for the matrix and fiber respectively. V_m and V_f are volume fractions of the matrix and fiber. The stress equilibrium and strain compatibility between the constituents are assumed to require

a) for equilibrium of constituent stresses,

$$\begin{aligned} d\sigma_m &= V_m\, d\sigma_x^m + V_f\, d\sigma_x^m \\ d\sigma_y &= d\sigma_y^m = d\sigma_y^f \\ d\tau_{xy} &= d\tau_{xy}^m = d\tau_{xy}^f \end{aligned} \tag{1}$$

b) for compatibility of constituent strains,

$$d\varepsilon_x = d\varepsilon_x^m = d\varepsilon_x^f$$

$$d\varepsilon_y = V_m \, d\varepsilon_y^m + V_f \, d\varepsilon_y^f \qquad (2)$$

$$d\gamma_{xy} = V_m \, d\gamma_{xy}^m + V_f \, d\gamma_{xy}^f \; .$$

We assume orthotropic elastic properties for the matrix, fiber and composite respectively, i.e., $d\underset{\sim}{\varepsilon} = \underset{\approx}{C}\, d\underset{\sim}{\sigma}$, $d\underset{\sim}{\varepsilon}^m = \underset{\approx}{M}\, d\underset{\sim}{\sigma}^m$, $d\underset{\sim}{\varepsilon}^f = \underset{\approx}{F}\, d\underset{\sim}{\sigma}^f$ where $\underset{\approx}{C}$, $\underset{\approx}{M}$, and $\underset{\approx}{F}$ are orthotropic elastic compliance matrices of the composite, matrix and fiber respectively. These matrices have non-zero elements C_{11}, C_{12}, C_{22}, C_{33} for $\underset{\approx}{C}$ and similar non-zero elements for $\underset{\approx}{M}$ and $\underset{\approx}{F}$ also.

Under a mechanical loading, the matrix is assumed to deform as an elastic-perfectly plastic material, i.e., $d\underset{\sim}{\varepsilon}^m = d\underset{\sim}{\varepsilon}^{me} + d\underset{\sim}{\varepsilon}^{mp}$, with the Prandtl-Reuss flow rule $d\underset{\sim}{\varepsilon}^{mp} = d\lambda(df^m/d\underset{\sim}{\sigma}^m)$ and the Von-Mises yield condition $f^m = (\sigma_x^m)^2 - (\sigma_x^m)(\sigma_y^m) + (\sigma_y^m)^2 + 3(\tau_{xy}^m)^2 - Y_m^2 = 0$ where $d\lambda$ is a proportionality constant, Y_m is the yield stress of the matrix. During a thermal change, the matrix may deform either elastically or plastically. Thus, if the matrix deforms elastically during thermal change, then $d\underset{\sim}{\varepsilon}^m = d\underset{\sim}{\varepsilon}^{me} + d\underset{\sim}{\varepsilon}^{mth}$, and if plastically then, $d\underset{\sim}{\varepsilon}^m = d\underset{\sim}{\varepsilon}^{mp} + d\underset{\sim}{\varepsilon}^{mth}$ where the superscripts e, p and th denote elastic, plastic and the thermal components, respectively. Note that during plastic deformation, an elastic component of the strain is absent because we assume a non-hardening plasticity for the matrix. External load is assumed to be constant during incremental change of the temperature. The elastic-plastic deformation caused by the external mechanical loading may be separated from that caused by the temperature change. The details of the analysis for the former may be found in Min and Min and Crossman[18-19]. For analysis of thermal deformation, here we assume that the initial internal stress state is completely known. Assuming that the orthotropic thermal expansion behavior of the matrix and the fiber (i.e., $d\underset{\sim}{\varepsilon}^{mth} = [\alpha_x^m, \alpha_y^m, o]dT$, $d\underset{\sim}{\varepsilon}^{fth} = [\alpha_x^f, \alpha_y^f, o]dT$) is known as a function of temperature, the coefficients of thermal expansion (CTE's) of the composite during the elastic deformation of the matrix can now be obtained,

$$\alpha_x^e = (V_f C_{11}\, \alpha_x^f + k_1 \, F_{11} \, \alpha_x^m)/(V_f \, C_{11} + k_1 \, F_{11}) \qquad (3)$$

$$\alpha_y^e = \frac{V_f\ C_{12} - V_f\ k_1\ F_{12}}{V_f\ C_{11} + k_1\ F_{11}}(\alpha_x^f - \alpha_x^m) + (V_m\ \alpha_y^m + V_f\ \alpha_y^f)$$

where $k_1 = 1 - C_f C_{11}/F_{11}$. If the relationships $C_{11} = k_1 M_{11}/V_m$, $C_{12} = k_1 M_{12}$ are substituted and the isotropic thermal expansion properties are assumed for the matrix and fiber, the above equations reduce to the equations given by, for example, Schapery[16]. If the matrix deforms plastically during thermal change, then the yield and the constraint condition Eq. (1) requires that no matrix stress increases during this change. We then have

$$\begin{aligned} d\varepsilon_x &= d\varepsilon_x^m = d\varepsilon_x^{mp} + \alpha_x^m dT = \alpha_x^f dT \\ d\varepsilon_y &= V_m(d\varepsilon_y^{mp} + \alpha_y^m dT) + V_f \alpha_y^f dT \\ d\gamma_{xy} &= V_m\ d\gamma_{xy}^{mp} \end{aligned} \tag{4}$$

By the first of the above equation, $d\lambda = [(\alpha_x^f - \alpha_x^m)/(3\sigma_x^{m'})]dT$, where the prime denotes a deviatoric component for the corresponding stress. The CTE's during the plastic deformation can then be calculated,

$$\begin{aligned} \alpha_x^p &= \alpha_x^f \\ \alpha_y^p &= V_m[(\alpha_x^f - \alpha_x^m)/\sigma_x^{m'}]\sigma_y^{m'} + (V_m\ \alpha_y^m + V_f\ \alpha_y^f) \\ \alpha_{xy}^p &= 2V_m(\alpha_x^f - \alpha_x^m)(\tau_{xy}^{m'}/\sigma_x^{m'}) \end{aligned} \tag{5}$$

Note that the thermal expansion of the metal matrix composites in the range where the matrix deforms plastically is not a pure material property, but depend on the residual stress and the external loading status. A predictive analysis for thermal expansion was performed numerically using the above results. Necessarily, the thermal analyses required a complete elastic-plastic deformation analysis with loading and unloading capability. Details of the elastic-plastic analysis and numerical procedures may be found in References 18 and 19.

CYCLIC THERMAL EXPANSION OF GR/AL AND GR/MG

MATERIAL

The Gr/Al and Gr/Mg studied here are composed of unidirectional composite core structure sandwiched between two unreinforced metal facings. The graphite fibers for both materials are Union Carbide high modulus VS0054. The aluminum in Gr/Al is 6061 alloy. A magnesium-aluminum-zinc alloy serves as the matrix for the Gr/Mg composite. Considered as unidirectional lamina, the Gr/Al and Gr/Mg plate has average volume fraction of 27.6% and 43.6%, respectively.

THERMAL EXPANSION MEASUREMENT

The use of strain gages for the measurement of thermal expansion and residual stresses were reported by various authors[20-26]. The methods utilize characteristic response of strain gages to temperature variations, known as 'apparent strain'. If the apparent strain of a strain gage bonded on a specimen material(S_1) and that of a same type gage on a reference material(S_2) are measured, then the thermal expansion of a specimen may be determined by the following formula,

$$\varepsilon_1(T) = [\varepsilon_1^A(T) - \varepsilon_2^A(T) + \varepsilon_2(T) - B]/C_1 \qquad (6)$$

with $C_1 = (1 - K\nu)/(1 - K\nu_0)$, where ε_1 and ε_2 are thermal strains of specimen and reference material respectively, ε_1^A and ε_2^A are apparent strains indicated by the strain gages bonded on the same materials respectively, B is a constant for zero adjustment, K is the transverse sensitivity of the strain gage, ν_1 is minus the ratio of thermal expansion of the specimen material in the direction perpendicular to the strain gage axis to that parallel to it, ν_0 is the Poisson's ratio of the manufacturer's standard material (ν_0 = 0.285 for the strain gages used in this work). The thermal response of a T50 PAN/F263 Graphite/Epoxy unidirectional composite in the longitudinal direction measured by the strain gage methods showed an agreement with those determined by the quartz dilatometry within the accuracy of $+0.2\ \mu\varepsilon/^{o}C$[25].

The strain gage transverse correction, repesented by C_1 in the above is particularly important for the measurement of longitudinal thermal expansion because of the larger thermal expansion of the these materials in the transverse direction than in the longitudinal direction. To reduce the transverse sensitivity effect, a strain gage with low transverse sensitivity factor is desirable. The type of strain gages used for the thermal expansion measurement of the Gr/Metal studied here are listed in Table 1. The particular strain gages were selected because of their availability in our stock. The strain gage selection may be

TABLE 1

	Gr/Al-T6 Longi-tudinal	Gr/Al-0 Longi-tudinal	Gr/Al-0 Trans-verse	Gr/Mg Longi-tudinal	Gr/Mg Trans-verse
Strain Gage Type	WK09-125 BT-350	CEA00-062 UW-350	WK09-125 BT-350	WK06-062 TT-350	WK06-062 TT-350
Gage Factor	2.08	2.15	2.08	2.02	1.91
Transverse Sensitivity	-0.024	0.007	-0.024	0.001	-0.009
Reference Material	Al6061 -T6	SRM 739	Al6061 -T6	SRM 739	SRM 739

TABLE 2

ELASTIC COMPLIANCES FOR THE COMPOSITE AND FIBER

$$
\begin{aligned}
C_{11} &= 0.428*10^{-4} & (0.295*10^{-7}),\\
C_{12} &= -0.136*10^{-4} & (-0.935*10^{-8}),\\
C_{22} &= 0.255*10^{-3} & (0.176*10^{-6}),\\
F_{11} &= 0.153*10^{-2} & (0.105*10^{-7}),\\
F_{12} &= -0.384*10^{-3} & (0.265*10^{-8}),\\
F_{22} &= 0.725*10^{-1} & (0.500*10^{-6})\ \ \mathrm{GPa}^{-1}\ (\mathrm{psi}^{-1});
\end{aligned}
$$

COEFFICIENT OF THERMAL EXPANSION FOR THE VS0054 GRAPHITE FIBER

$$\alpha_x^f = -0.36(-0.2),$$

$$\alpha_y^f = 18.0(10.0)\ \mu m/m/^oC(\mu in/in/^oF);$$

THE EQUATION FOR LINEAR THERMAL EXPANSION OF 6061-T6 ALUMINUM

$$
\begin{aligned}
\varepsilon_x^m = \varepsilon_y^m &= -0.1105*10^{-5} + 0.1193*10^{-8}T^2\\
&\quad +0.4490*10^{-8}T^2 - 0.4071*10^{-11}T^3\ \mathrm{in/in}\,(^oF)
\end{aligned}
$$

based on the apparent strain repeatability that depends on the alloy composition and the geometrical shape. A close match between the CTE's of the strain gage alloy and of the specimen is also desired to increase the accuracy of the measurement. Two reference materials were used: 6061-T6[27] and fused silica-NBS Standard Reference Material (SRM) 739[28].

RESULTS AND DISCUSSION

GR/AL

Fig. 1 shows the thermal expansion of VS0054/6061-T6 Gr/Al in the longitudinal direction. Within the precision of the X-Y recorder output, appreciable variation from cycle to cycle was not observed during the second half of the first to sixth cycles. The information during the first half cycle was not recorded because this was the curing cycle for the strain gage bonding agent. The solid line in the figure is the curve predicted by assuming the initial residual stress of 106 MPa(15.4 ksi) and the yield stress of the matrix 276 MPa(40 ksi). Other material properties used for the predictive calculations are listed in Table 2. The thermal expansion follows the elastic prediction (Eq. 3) with fairly good accuracy. No plastic behavior is observed for this material for the temperature range considered. To see the effect of heat

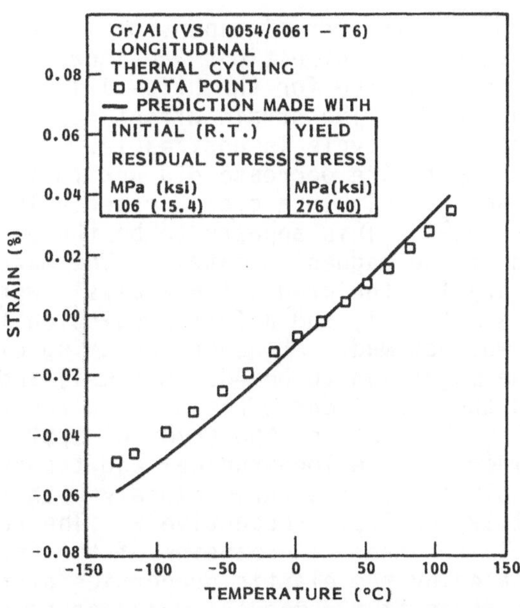

Fig. 1 Cyclic thermal expansion (5 cycles) of VS0054/6061-T6 Gr/Al unidirectional composites in longitudinal direction

treatment on the thermal expansion behavior, the specimens were annealed at 482°C for 30 minutes and air-cooled. The results of the thermal cycling for this specimen are shown in Fig. 2(a) for the longitudinal behavior. The square dots are the experimental points and the solid line is the one predicted with both the assumed initial residual and yield stresses of 39.8 MPa. Since the strain gage used here was not rated for -73.3°C, the data points below this temperature were discarded. Data above this temperature were repeatable from cycle to cycle within the measurement error.

The predicted curves show that the specimen exhibits thermo-elastic and thermo-plastic behavior during the thermal cycling, resulting in the characteristic hysteresis loop. The observed behavior does not delineate these regions as clearly as the theoretical one does. The smooth transition of the observed behavior from the elastic to plastic region appears to be due to the inhomogeneous yielding of the matrix on a micromechanistic scale and/or to the effect of creep. Fig. 2(b) shows corresponding results for the transverse direction. The agreement between the prediction and the measurement is quite reasonable in this direction. Fig. 2(c) shows the theoretical 'residual stress' change during this thermal cycling.

GR/MG

The results of cyclic thermal expansion for Gr/Mg in longitudinal and transverse directions are shown in Fig.3. The thermal strains were measured for the R.T.→ 121°C→ -73.3°C→ 121°C→ -73.3°C temperature path. The strain offset during the second cycle following the first cycle is noticeable. The thermal strain curve during the temperature decrease did not follow that during the temperature increase within a cycle, especially at temperatures above 38°C. This appears to be the effect of the creep deformation of the magnesium matrix. The material properties necessary for the predictive analysis were not available for this material, and a full prediction for this thermal behavior was not made. However, assuming elastic modulus and the CTE of the magnesium to be 45.5 MPa(ksi) and 26.1 μm/m/°C respectively, and the same fiber CTE's as used for the Gr/Al analysis in Eq.(3), the CTE's of the Gr/Mg of 1.82 and 27.7 μm/m/°C are obtained for the longitudinal and transverse directions respectively. These agree closely with the measured average values, 1.39 and 27.2, respectively. The results also indicate that while the thermal expansion of the Gr/Mg is essentially dominated by the elastic properties of the constituents, the creep of the magnesium causes some accumulation of the thermal strain during the thermal cycling. The last point may require further investigation because the present study provides only a limited amount of data on rate dependence.

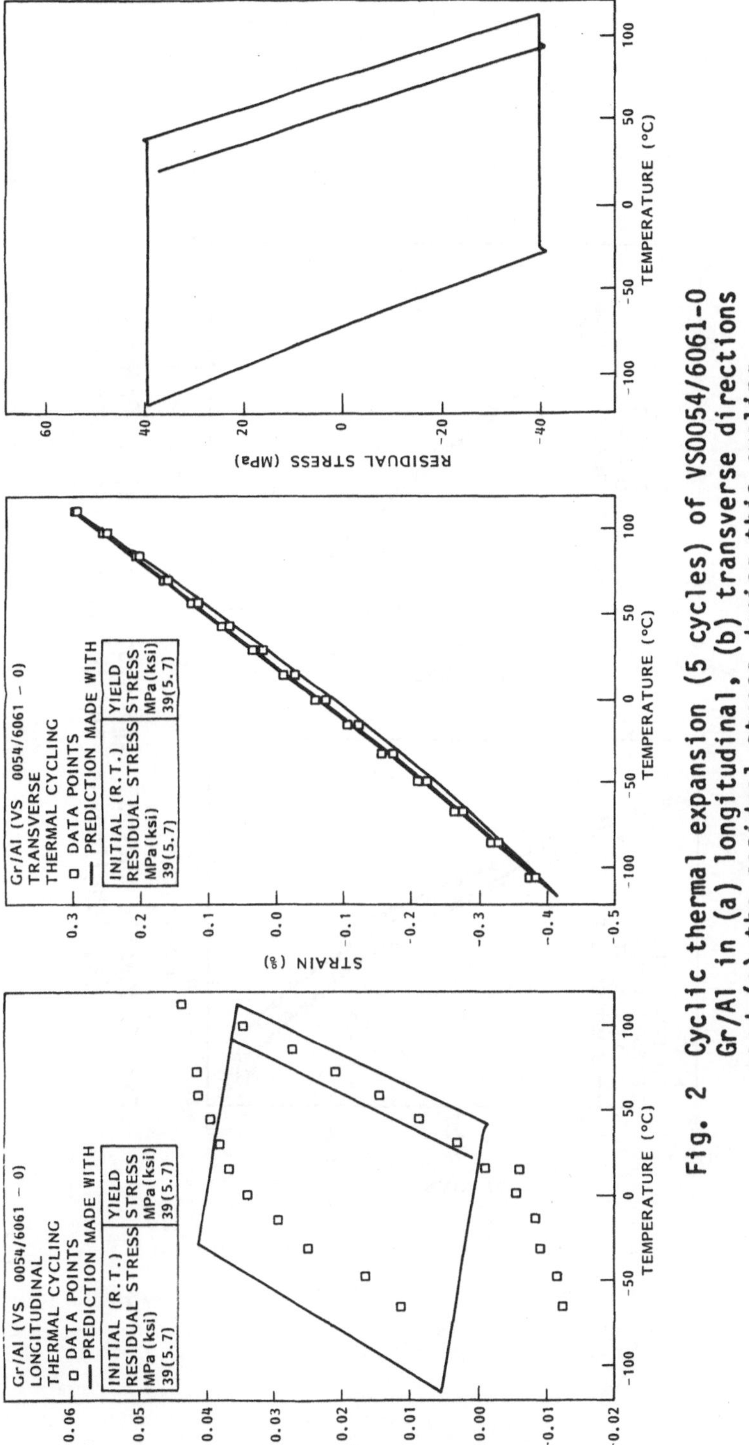

Fig. 2 Cyclic thermal expansion (5 cycles) of VS0054/6061-0 Gr/Al in (a) longitudinal, (b) transverse directions and (c) the residual stress during this cycling.

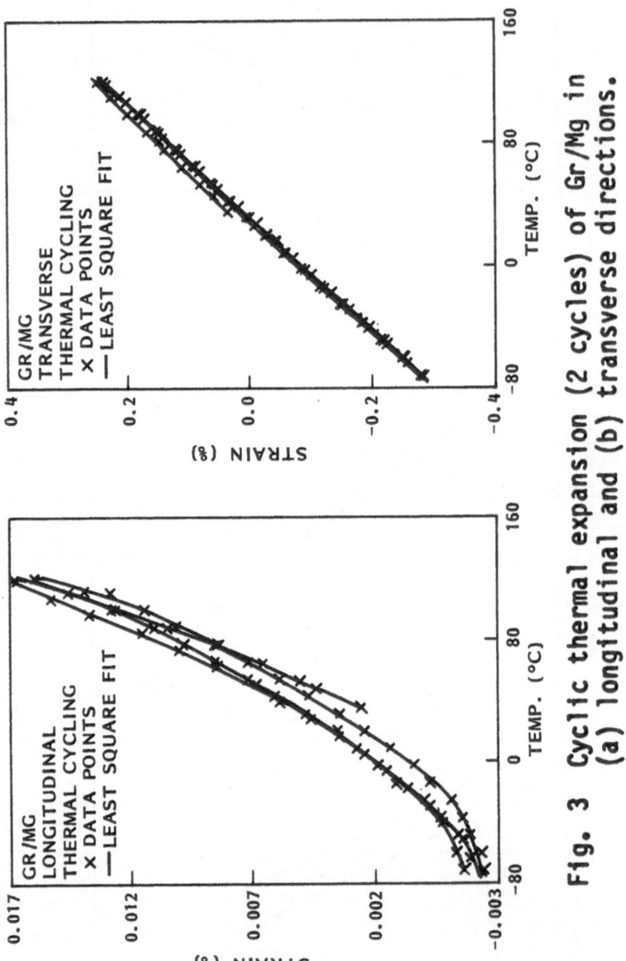

Fig. 3 Cyclic thermal expansion (2 cycles) of Gr/Mg in (a) longitudinal and (b) transverse directions.

SUMMARY AND CONCLUSION

Cyclic thermal expansion of Graphite/Aluminum and Graphite/Magnesium are measured by the strain gage techniques. The thermal expansion properties of the continuous fiber reinforced metal matrix composites are found to strongly depend on the elastic-plastic and creep behavior of the matrix metals and on the initial residual stresses. These effects are analyzed by a semi-empirical continuum model that accounts for the thermally induced elastic and plastic deformation of the matrix; the predicted results are shown to reasonably well correlate with the experimental data.

ACKNOWLEDGMENT

This work was performed under Lockheed Independent Research and Development program.

REFERENCES

1. Kreider, K. G. and Patarini, V. M., Met. Trans. 1, pp. 3431-3435 (1970).

2. Ferte, J.P., "La Dilatation Thermique des Materiaux Composites a Matrice Metallique: Analysis Experimentale et Interpretation Theorique," Ph.D thesis, Lyon (1974).

3. Wakashima, K., Kawakabe, T. and Umekawa, S., Met. Trans. A, 6A, pp. 1755-1760 (1975).

4. Larsson, L.O.K., ICCM 2, Toronto, Canada, pp. 805-821 (1978).

5. Umekawa, S, Wakashima, K, and Yoda, S., ICCM 2, Toronto, Canada, pp. 828-839 (1978).

6. Ishikawa, T, Koyama, K., and Kobayashi, S., J. Comp. Mat., Vol. 12, p. 153 (1978).

7. Adams, D. F., Doner, D. R., and Thomas, R. L., "Mechanical Behavior of Fiber Reinforced Composite Materials," AFML-TR-67-96(AD-654065) (1967).

8. Greszczuk, L.B., AIAA 6th Str. and Mater. Conf., Palm Springs, Calif., p. 285 (1965).

9. Dvorak, G. J. and Rao, M. S. M., J. Appl. Mech., Dec. p. 619 (1976).

10. Crossman, F. W. and Karlak, R. F., Failure Modes in Composites II, J. N. Fleck and R. L. Mehan, ed. AIME, pp. 8-31 (1974).

11. Foye, R. L., J. Comp. Mat., Vol. 7, p. 148 (1973).

12. Adams, D.F., J. Comp. Mat., Vol. 4, p. 310 (1970)

13. Weiss, H. J., J. Mat. Sci. 12, p. 797 (1977).

14. Garmong, G., Met. Trans. Vol.5, p. 2183, (1974).

15. de Silva, A. R. T. and Chadwick, G. A., J. Mech. Phys. Solids, Vol. 17, p. 387 (1969).

16. Schapery, R. A., J. Comp. Mat., Vol. 2, No. 3, pp. 380-404 (1968).

17. Bahei-El-Din, A. and Dvorak, D. J., Modern Development in Composite Materials and Structures, J. R. Vinson, ed., ASME, pp. 123-147 (1979).

18. Min, B.K., "A Plane Stress Formulation for Elastic-Plastic Deformation of Unidirectional Composites," J. Mech. Phys. Solids, Vol. 29, NO. 4, pp. 327-352 (1981)

19. Min, B. K. and Crossman, F. W., "History-Dependent Thermo-Mechanical Behavior of Unidirectional Graphite-Aluminum Composites," Sixth Conference of Composites: Testing and Design, ASTM, Phoenix, AZ (1981).

20. Wang, A. S. D., Pipes, R. B., and Ahmadi, A., Composite Reliability, ASTM STP 580, p. 574 (1974).

21. Daniel, I. D., Thermal Expansion 6, I.D. Peggs, ed., p. 203 (1978).

22. Tennyson, R. C., 21st Structures, Structural Dynamics and Materials Conference, Part I, pp. 1009-1018, 1980, American Institute of Aeronautics and Astronomy.

23. Bowles, D. E., Post, D., Herakovich, C. T., and Tenney, D. R., "Thermal Expansion of Composites Using Moire Interferometry," VPI-E-80-19 (1980).

24. Wolff, E. G. and Susskind, C. S., "The Measurement of Dimensional Stability of Invar Using Strain-Gage Techniques," SAMSO-TR-76-186 (1976).

25. Min, B. K. and Bartanen, B. W., "Cyclic Thermal Expansion Measurement of Unidirectional Composites by Strain Gage Methods," LMSC-D766717 (1980).

26. Diver, G. and Pavlovic, A.S., "Thermal Expansion of Polycrystalline ZnF ," Eighth International Thermal Expansion Symposium, Gaithersburg, Md., this volume (1981).

27. ALCOA Aluminum Handbook, the Aluminum Company of America, p. 43 (1967).

28. Hahn, T. A. and Kirby, R. K., Thermal Expansion-1971, M. G. Graham and H. E. Hagy, ed. p.13, (1971).

THERMAL EXPANSION OF BINARY CoAl, FeAl, AND NiAl ALLOYS

R. W. Clark*

J. Daniel Whittenberger**

* Department of Physics and Engineering
Pacific Lutheran University
Tacoma, WA 98447

** Materials Division
NASA-Lewis Research Center
Cleveland, OH 44135

ABSTRACT

As part of a program to study intermetallic alloys as potential high temperature structural materials for use in severe environments, the thermal expansion of binary CoAl, FeAl, and NiAl was measured as a function of composition and temperature. Approximately eight alloys covering a range of about eight atom percent were examined for each system from room temperature to 1273 K. The thermal expansion of all three intermetallics varies only slightly with composition. The expansion of CoAl is similar to, but somewhat less than, that of NiAl. The measured expansion of NiAl is identical with the Thermophysical Properties Research Center (TPRC) provisional curve. While the measured thermal expansion of FeAl differs considerably from the TPRC provisional values, this study agrees with Ho and Dodd's results for FeAl (Scripta Met., 12 (1978), 1055).

INTRODUCTION

A program has been initiated at the Lewis Research Center to investigate the potential of intermetallics for use as high temperature structural materials in severe environments. Due to their high melting points, cubic crystal structure, extensive solid solubility, and excellent oxidation resistance; the binary CoAl, FeAl, and NiAl systems were selected for the initial studies.

While technologically important for design purposes alone, knowledge of the thermal expansion of these aluminides is also necessary to develop elevated temperature deformation models. These data and lattice parameters as a function of temperature and composition are required to calculate vacancy concentrations. CoAl, FeAl, and NiAl have complex point defect structures consisting of antistructure (substitutional) atoms and vacancies (refs. 1-3). The number and type of defects are dependent on overall alloy composition as well as temperature. In comparison to metals these aluminides can contain enormous numbers of vacant atom sites -- approaching 10 percent for Al-rich alloys (refs. 3-5). Such large concentrations of vacancies could have a profound influence on mechanical properties.

To our knowledge no studies of the thermal expansion of CoAl have been undertaken. Only a limited amount of thermal expansion data exists for the FeAl and NiAl systems. Ryabov, et al (ref. 6) measured the expansion of a 50.3 at pct Fe - 49.7 pct Al alloy, and Ho and Dodd (ref. 7) have made thermal expansion measurements on five binary FeAl alloys ranging in composition from 51 to 45.5 pct Al between room temperature and approximately 1175 K. Examination of this work reveals that Ho and Dodd observed more expansion at the higher temperatures than Ryabov et al, and that thermal expansion is a slight function of alloy chemistry. Two studies of the expansion of NiAl have also been conducted. Wasilewski (ref. 8) determined the thermal expansion of several NiAl alloys ranging in composition from 48.2 to 51.4 pct Al between 4 and 1273 K. In opposition to Ho and Dodd's finding for FeAl, Wasilewski reports that the thermal expansion of NiAl is not dependent on composition. Singleton, et al (ref. 9) have also measured the expansivity of a NiAl alloy (unknown composition) from 266 to 1447 K; however they measured somewhat less expansion than Wasilewski at all temperatures.

This paper describes a study of the thermal expansion of CoAl, FeAl, and NiAl alloys as a function of composition and temperature from room temperature to 1273 K. About eight alloys covering a range of eight atomic percent for each system were fabricated via powder metallurgy techniques. Absolute thermal expansion of the various alloys was determined from the relative expansion for each alloy and that for a Pt standard.

EXPERIMENTAL PROCEDURES

Because of the difficulty of producing sound material by traditional melting and casting techniques, all the intermetallic alloys were produced by powder metallurgy methods, as outlined in Fig. 1. Lots of prealloyed powder with Al contents near the maximum and minimum solubility limits for each system were commercially procured. Based on the certified compositions supplied with the powders, six or seven intermediate compositions were chosen and produced for each system by powder blending techniques. The chemistries for the alloys are listed in Table I. Since the Al atom/metal atom ratio equal to 1.0 is the boundary between predominate point defect types, the intermediate compositions were biased toward the equiatomic value. While this protocol was successful for CoAl and NiAl, an incorrect analysis of the Al-rich FeAl alloy powder produced an inconsistent set of intermediate alloy compositions in the FeAl series.

Fully dense intermetallic alloys were produced by hot extrusion techniques. The alloy powders were placed in mild steel extrusion cans which were then sealed under vacuum after being heated to about 500 K to drive off any possible water. The cans were introduced into a furnace at 1450 (NiAl) or 1505 K (CoAl and FeAl), held at temperature for at least one hour and then extruded into round bar at a 16:1 reduction ratio. The CoAl alloys were

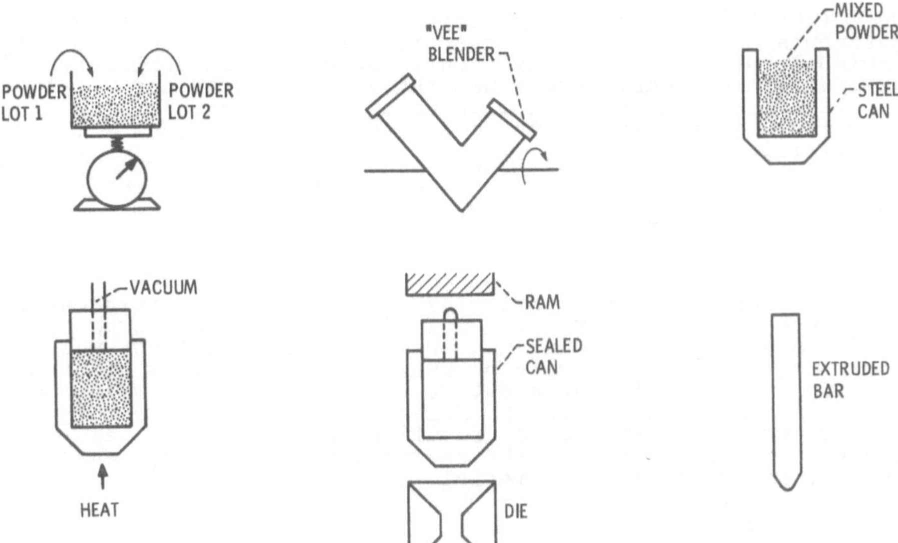

Figure 1. Schematic outline of powder metallurgical techniques utilized to produce intermetallic alloys.

considerably stronger than either FeAl or NiAl and required essentially the full available pressure of 1300 MPa to begin extrusion as opposed to nominally 650 MPa for the other systems.

Nominally 15 mm long, 5 mm diameter specimens were electrodischarge machined from extruded bars of each composition. The relative expansion of such specimens was determined as a function of temperature from room temperature to about 1300 K through use of a horizontally mounted alumina rod and tube dilatometer where a LVDT measured the differential motion and a

Table I. Composition of Intermetallic Alloys. *

Alloy Identification		CoAl	FeAl	NiAl
		Aluminum, Atomic Percent		
CoAl- FeAl- NiAl-	1	51.31	48.71	52.67
	2	50.5	48.19	51.6
	3	49.89	47.48	50.6
	4	49.3	46.76	49.91
	5	48.5	45.66	49.2
	6	47.5	44.55	48.25
	7	46.5	43.24	47.0
	8	44.92	41.72	45.5
	9			43.94

* All alloys contained nominally .02 C, .05 H, .004 N, and .10 O.

type K (chromel/alumel) thermocouple recorded the temperature. All experiments were conducted at heating rates of 3 K/min or less; individual specimens were rerun until agreement was obtained among 2 or 3 determinations. Published thermal expansion data (ref. 10) in combination with our relative data for a Pt standard were used to calculate absolute thermal expansion values for the intermetallic alloys.

RESULTS and DISCUSSION

Materials

Metallographic examination of the as-extruded intermetallic alloys revealed that they were, for all practical purposes, fully dense; since porosity was not observed. All alloys possessed uniform, equiaxed grains with nominal grain diameters of about 15 µm for CoAl and NiAl and about 30 µm for FeAl. Typical grain structures are shown in Fig. 2. Selected alloys were examined for preferred crystallographic orientation; however none was found. Therefore it was felt that the extruded alloys are representative of sound, polycrystalline material.

Thermal Expansion

The relative expansion for each alloy was combined with the system correction factor, determined from 18 measurements of a Pt standard, to obtain thermal expansion. Examination of these data revealed that the expansivity is not a strong function of composition for these three aluminides within the temperature range investigated. Therefore all data were joined for each system (160 points for CoAl and FeAl and 180 for NiAl) and fitted as a function of temperature by means of computerized linear regression techniques. The results of these fits are presented in Fig. 3, and the pertinent equations are given below:

$$\text{CoAl} \quad \Delta\ell/\ell_0 = -0.00164 + 1.072\times10^{-3}(T-293) + 3.98\times10^{-7}(T-293)^2,$$

$$\text{FeAl} \quad \Delta\ell/\ell_0 = -0.00509 + 1.739\times10^{-3}(T-293) + 2.2\times10^{-8}(T-293)^2 + 7.39\times10^{-10}(T-293)^3,$$

$$\text{NiAl} \quad \Delta\ell/\ell_0 = -0.00091 + 1.281\times10^{-3}(T-293) + 2.70\times10^{-7}(T-293)^2 - 3.62\times10^{-11}(T-293)^3;$$

where $\Delta\ell/\ell_0$ is the thermal expansion, and T is the temperature in K. The regression coefficients for these equationsare .998 or better, and the standard deviations of fit about the predicted $\Delta\ell/\ell_0$ values are .0174 for CoAl and NiAl and .0255 for FeAl.

The curves in Fig. 3 illustrate the significantly larger thermal expansion of FeAl in comparison to that of either CoAl or NiAl. Also the data reveal that the expansion of NiAl is always slightly greater than that of CoAl between room temperature and 1273 K. However the difference at any temperature is very small, less than 0.08 pct.

The data for FeAl and NiAl in Fig. 3 were compared to the literature and Thermophysical Properties Research Center (TPRC) provisional curves((ref. 11). In the case of NiAl the TPRC curve is based solely on Wasilewski's results (ref. 8), and our data are in total agreement with this estimate; in fact the agreement is quite remarkable with a maximun deviation of 0.02 pct at 1273 K.

Figure 2. Typical photomicrographs of intermetallic alloys. The extrusion axis is vertical.

The thermal expansion for FeAl is, however, another matter. Fig. 4 compares the TPRC provisional curve (ref. 12) with the expansion data of Ho and Dodd (ref. 7) and that from this work. The TPRC curve is founded on Ryabov et al's data (ref. 6), and the plot describing Ho and Dodd's study is the result of our regression fit of their data from alloys with 47. to 51. at pct Al (regression coefficient = .997; standard deviation about the predicted $\Delta\ell/\ell_0$ = .036). While there is general agreement below 773 K among the

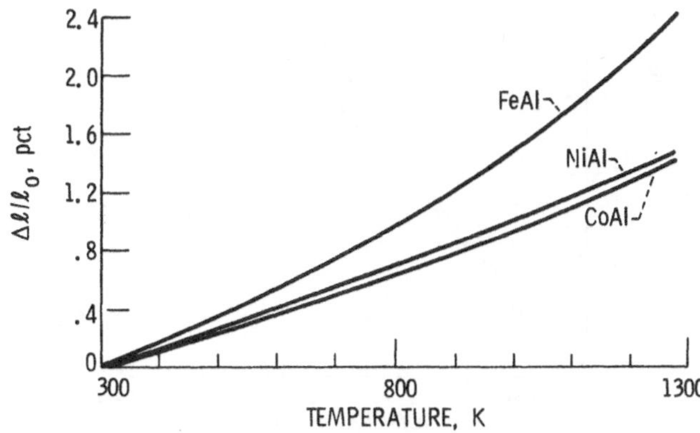

Figure 3. Thermal expansion of CoAl, FeAl and NiAl as a function of temperature.

three sets of data, the behavior shown in Fig. 4 demonstrates that significantly more expansion occurs in FeAl than was previously believed (ref. 12). On the other hand the agreement between this study and that of Ho and Dodd is excellent at all temperatures where the maximun difference is 0.03 pct at 1173 K.

Although variations in thermal expansion due to chemistry were ignored in the above, a small dependency on alloy composition was found in these aluminides. This is illustrated in Fig. 5 where the thermal expansion of alloys which possessed the maximum and minimum expansivity is plotted. CoAl (Fig. 5(a)) is well behaved with the amount of expansion decreasing with increasing Al content. The thermal expansion of NiAl (Fig. 5(c)) is regular, but it first increases as the composition increases from 44 pct to 47 pct Al and then decreases with increasing Al. The behavior of FeAl (Fig. 5(b))is not so simple; here opposite trends with respect to composition are observed. Below 1100 K the Al-rich alloys show less expansion that the Al-poor alloys; however above 1100 K the amount of thermal expansion increases with increasing Al content.

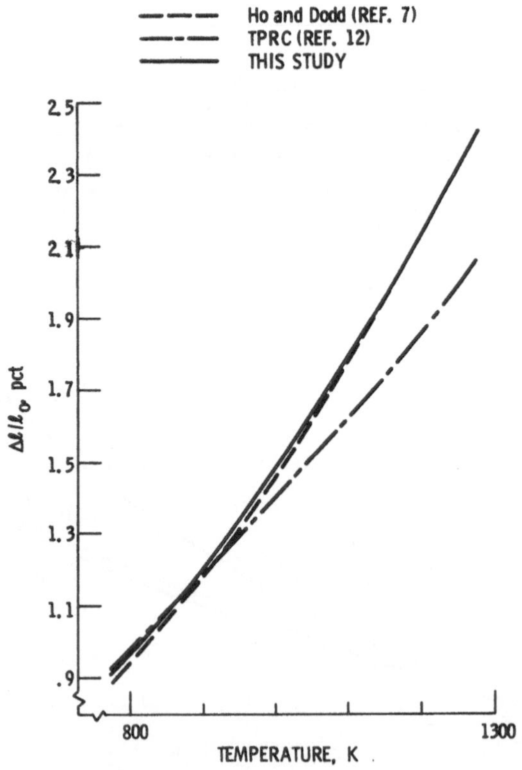

Figure 4. Thermal expansion of FeAl as measured by several investigators.

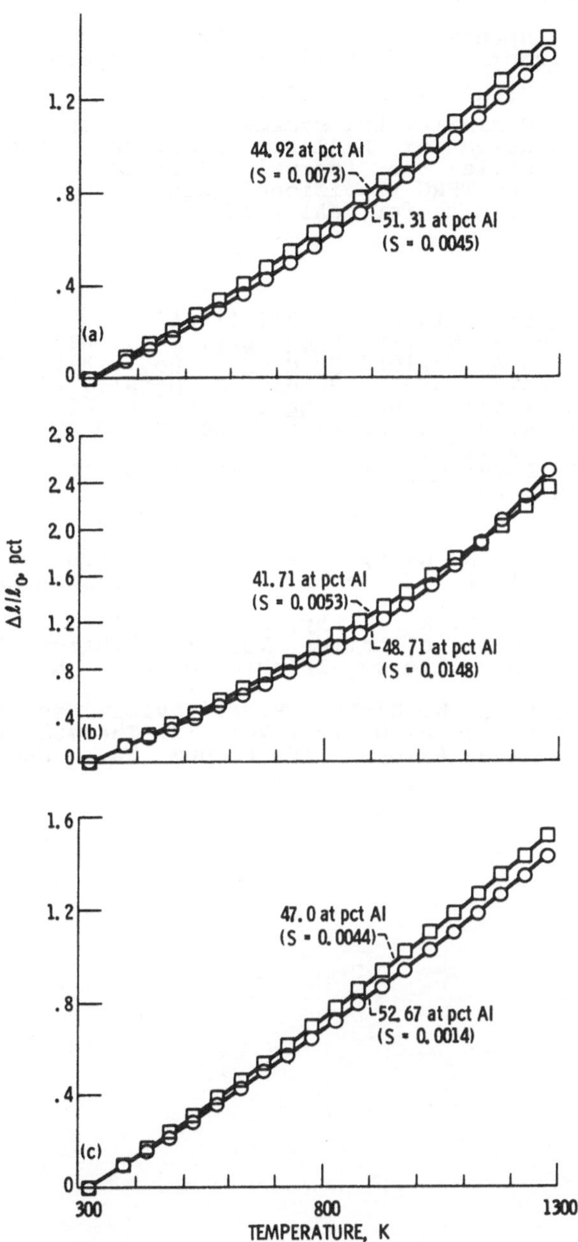

Figure 5. Thermal expansion of (a) CoAl, (b) FeAl and (c) NiAl as functions of composition and temperature. Maximum range in data for expansion was 0.026 pct for CoAl and 0.05 pct for FeAl and NiAl. Regression analysis of all data yielded Regression coefficents > 0.9995; standard deviation about predicted $\Delta\ell/\ell_0$ values, S, is given for each curve.

SUMMARY of RESULTS

The thermal expansion of CoAl, FeAl, and NiAl has been investigated as functions of chemistry and temperature from room temperature to 1273K. The thermal expansion of all three intermetallics varies only slightly with composition. The expansion of CoAl is similar but somewhat less than that of NiAl. The measured expansion of NiAl is identical with the TPRC provisional curve. While the thermal expansion of FeAl differs considerably from the TPRC provisional values, this study agrees with Ho and Dodd's results for FeAl (ref. 7).

REFERENCES

1. N. Ridley: J. Inst. Metals, 94, 255 (1966).
2. A. J. Bradley and A. H. Jay: Proc. Roy. Soc., A136, 210 (1932).
3. A. J. Bradley and A. Taylor: Proc. Roy. Soc., A159, 56 (1937).
4. R. Meyer, E. Wachtel, and V. Gerold: Z. Metal., 67, 97 (1976).
5. D. Paris, P. Lesbats: J. Nucl. Mater., 69-70, 628 (1978).
6. V. R. Ryabov, A. V. Lozovskaya, and V. G. Vasilyev: Phys. Metals Metal., 27, 98 (1969).
7. K. Ho and R. A. Dodd: Scripta Met., 12, 1055 (1978).
8. R. J. Wasilewski: Acta Met., 15, 1757 (1967).
9. R. H. Singleton, A. V. Wallace, and D. G. Miller: AFML-TR-66-197, (1966).
10. T. A. Hahn and R. K. Kirby: AIP Conference Proceedings, No. 3 Thermal Expansion - 1971 (Third Symposium), American Institute of Physics, New York, 87 (1972).
11. Y. S. Touloukian, R. K. Kirby, R. E. Taylor, and T. Y. Lee: Thermophysical Properties of Matter, Vol. 12, Thermal Expansion - Metallic Elements and Alloys, IFI/Plenum Co., New York, 438 (1977).
12. Y. S. Touloukian, R. K. Kirby, R. E. Taylor, and T. Y. Lee: Thermophysical Properties of Matter, Vol. 12, Thermal Expansion - Metallic Elements and Alloys, IFI/Plenum Co., New York, 433 (1977).

MATERIAL PROCESSING AND THERMAL EXPANSION PERFORMANCE OF THREE INVAR ALLOYS

Timothy P. O'Donnell
Wallace M. Rowe

Applied Mechanics Technology Section
Jet Propulsion Laboratory
California Institute of Technology
Pasadena, California

INTRODUCTION

With the evolution of longer mission, more sophisticated spacecraft and their associated scientific instruments, increasing demands have been placed on materials of construction to maintain both short and long term dimensional stability. The Wide Field/Planetary Camera (WF/PC), being designed and built by Jet Propulsion Laboratory/California Institute of Technology, is an example of this type of high technology science instrument. The WF/PC will occupy one bay of the NASA Space Telescope which is scheduled for shuttle launch (1985). This instrument contains two complete optical relay and detector systems. One relay system will be operated at f/12.9 for wide-field work, and the other at f/30 for planetary and other higher resolution work. Experiments planned for the WF/PC require geometric stability in the detectors to a fraction of a picture element (pixel) over long exposures. Thus, materials and processes used for the Optical Bench Structure of this instrument had to be selected judiciously to assure an extremely stable assembly. In past spacecraft programs, e.g., Viking and most recently Voyager, invar alloys have enjoyed much success in similar applications involving optical assemblies for imaging science instruments. Therefore, they were prime candidates for the WF/PC along with the relatively new (for this type application) graphite/epoxy composites. While both materials are being used on the instrument, this paper will address only different invar alloy applications and the effort expended to characterize their thermal expansion behavior. Other information on WF/PC optical bench materials can be found elsewhere(1).

Table 1. Chemical Composition of Selected Invar Alloys

ALLOY	C	P	S	SI	Mn	Se	Ni	Fe
Low-C Invar*	0.030	0.010	0.010	0.35	0.40	-	35.92	bal
Low-C Invar**	0.031	0.005	0.011	0.36	0.43	-	35.3	bal
Invar 36	0.15 max	-	-	0.35 max	0.50 max	0.15-0.25	35-36.5	bal
LR-35***	0.06	0.004	0.003	0.02	<0.01	-	36.05	bal

* hot forged, 1450°F anneal, Simonds Steel Heat No. 5292
** Boeing Analysis, same lot of material as *
*** Unispan by Universal Cyclops

Three invar alloys were investigated, representing various levels of carbon/alloying element content. Alloy elements affect machineability, thermal expansion and dimensional stability. Carbon content, for example, has been found to be very instrumental in affecting the dimensional behavior of invar alloys (2). Reasons for their selection ranged from what was readily available and in stock (free machining Invar 36) to material representative of the latest technology development in long term dimensional stability properties. Super Invar (Fe-Ni-Co) was considered for use at one point. Dimensional stability has been reported (3) to be $0 \pm 0.03 \times 10^{-9}$ over a 170 day period at $7^{o}C$. However, due to concerns over low temperature (about $-50^{o}C$) phase change with corresponding radical changes of coefficient of thermal expansion (CTE) (4) and our lack of experience with this alloy, it was not used.

Samples prepared from the three alloys were subjected to selected material processing sequences which represented current practices at JPL for thermal treatment and machining operations on invar alloys. After the samples were given accelerated cycling treatments to simulate long term service conditions, their thermal expansion characteristics were measured. Special attention was given to material processing sequences to assure elimination of residual stresses and thereby, the acquisition of long-term dimensional stability.

SPECIMEN PREPARATION

Material Description and Specimen Fabrication

Three invar alloys were selected for this investigation. Their choice was based on desired characteristics for various instrument components such as good machineability, low thermal expansion coefficient and long-term dimensional stability. The chemical composition of these materials is shown in Table 1. It should be noted that actual analyses were performed on the Low-Carbon (Low-C) and the LR-35 invars, whereas the Invar 36 values are from the military specification. The differences in the three alloys are primarily in their carbon (C) content. The Invar 36 came from stock that was on hand, residual from past spacecraft programs. The Unispan LR-35 was residual from the Viking Program. The Low-C invar was purchased from Simonds Steel in the hot forged, then annealed condition for the WF/PC project. Low C billet sizes of 2.75 in. by 15.5 in. by 24.5 in. were saw cut prior to delivery to JPL. Material test certifications were provided with this shipment.

Austenitic invar alloys have a high capacity for plastic and associated elastic deformation during machining operations. These alloys are tough and ductile and, therefore, very susceptible to

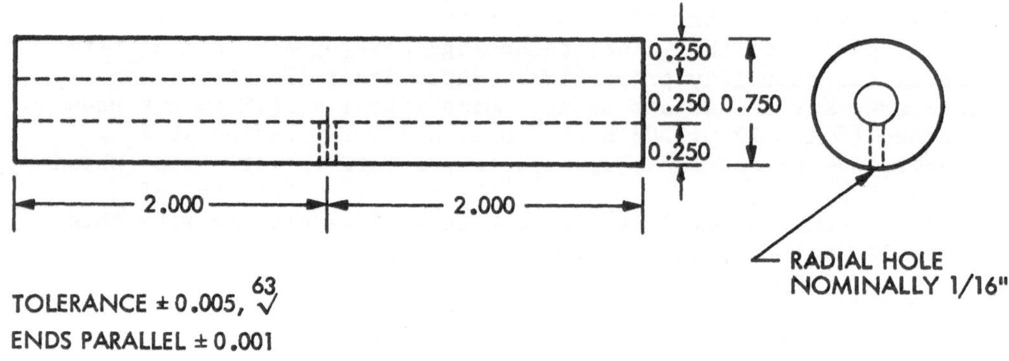

UNIVERSITY OF ARIZONA SAMPLE DIAGRAM

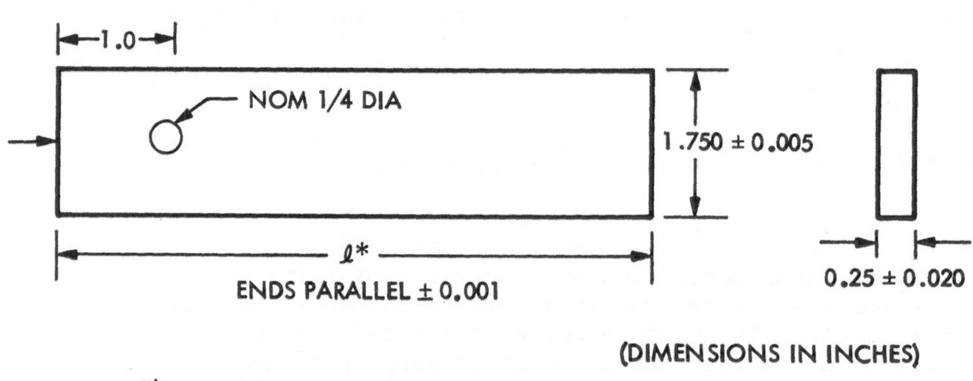

COMPOSITE OPTICS SAMPLE DIAGRAM

Fig. 1. Examples of thermal expansion test sample dimensions.

residual stress buildup. To deal with the above phenomena, care was taken in machining specimens to assure minimization of residual stresses. This was accomplished by incorporating thermal annealing and stress relieving treatments (which are described later) between machining steps.

Specimens were prepared for thermal expansion tests to conform to the configuration requirements of each particular measurement technique. For example, tests performed at the University of Arizona required machined, annular cylinders while those tests performed at Composite Optics Inc. required a flat, rectangular bar configuration. Figure 1 illustrates the configuration of these specimens.

Specimen Treatment

Various heat treatments were performed on the samples depending on what combination of properties was desired. All heat treating operations were performed in furnaces that conformed to MIL-H-6875 requirements for class B, C, and D Steels. A description of these heat treatments is as follows: Where it was desired that the finished part have an optimum combination of low thermal expansion and dimensional stability with minimal part warpage during quench operations the following thermal cycle was done:

OIL QUENCH

1. $1525^{o}F \pm 25^{o}F$, 1/2-1 Hr, Agitated Oil Quench
2. $600^{o}F \pm 20^{o}F$, 1 Hr, Air Cool
3. $200^{o}F \pm 5^{o}F$ 48 Hr, Air Cool

For optimum long term dimensional stability, relatively low coefficient of expansion and essentially no part warpage, the following thermal treatment was performed:

ANNEAL

1. $1450^{o}F \pm 25^{o}F$, 1/2-1 Hr, Furnace Cool to $600^{o}F$ ($\leq 200^{o}F$/Hr) then air cool
2. $600^{o}F \pm 20^{o}F$, 1 Hr, Air Cool
3. $200^{o}F \pm 5^{o}F$, 48 Hr, Air Cool

For some parts, where 1) the lowest coefficient of expansion has been desired, 2) long-term stability was not a big concern and 3) part warpage from heat treatment was not a problem, the following thermal cycle has been used:

WATER QUENCH

1. 1525°F ± 25°F, 1/2-1 Hr, Agitated water quench
2. 600°F, 1 Hr, Air Cool
3. 200°F ± 5°F, 48 Hr, Air Cool

Careful incorporation of various machining operations with the different thermal treatment cycles is of utmost importance. Dimensional stability is at stake (4 and 5). We have found that different thermal/mechanical procedures for different configured invar parts is not unusual.

The general technique used in preparing thermal expansion samples for testing in this characterization program was as follows: 1) Rough machine (taking progressively lighter cuts) leaving 0.020 to 0.030 inch, 2) Elevated temperature treatment, 3) Finish machine (final cuts $\leq$ 0.005 inch), 4) Stress relieve in inert atmosphere at 600°F, 5) Low temperature age, and 6) Handle with care. Techniques used to prepare flight parts were not necessarily the same as above. Consideration was given to the various cross sectional areas needed for final dimensions. Thinner parts, < 0.030 inch, are much more affected by any imparted surface residual stress. During early developmental machining work on these type parts warping was observed, sometimes after machining and at other times after heat treat operations. In some cases this problem was solved by machining to finish dimensions then "jigging" the part in place and going through the heat treat operations utilizing inert atmospheres. These tough and ductile alloys are, thus, very susceptible to the residual stress phenomenon. Thinner machined parts can possess higher overall stress levels and warping can and has been observed.

TESTS

Thermal expansion measurements on invar alloys have been made at Boeing, General Dynamics - Convair Division, Composite Optics Inc., University of Arizona and Simonds Steel Co. for JPL. In this program, ten tests were done on JPL prepared "coupon" samples, four measurements were made on JPL-prepared flight parts, three on Boeing prepared coupons and one on a Simond Steel Co. prepared sample.

Measurement Techniques

Relative length changes of samples were made at Composite Optic Inc., San Diego, California utilizing a unique optic comparator. This laser optic strain gage measured the relative length change between the provided sample and a ULE standard of known expansion. Dilatometer type measurements were made at Simond Steel and Boeing.

Actual length changes of samples were made at University of Arizona Optical Science Center in Tuscon, Arizona; Boeing Company in Seattle, Washington and General Dynamics-Convair Division, San Diego, California. The Univerisity of Arizona utilized the dependence of a Fabry-Perot etalon's resonanant frequency on mirror separation. The sample was formed into a spacer with a highly reflecting endplate optically contacted to each end. Length changes were monitored by means of a variable frequency He-Ne laser. Changes in frequencies then allowed the computation of the coefficient of thermal expansion α through the relationship

$$\alpha = (1/\Delta T)(\Delta L/L) = (1/\Delta T)(\Delta V/V) \qquad (1)$$

where v is the stable laser frequency and ΔL is the length change over a temperature interval ΔT for a sample length L. This technique reportedly gives an accuracy of $1 \times 10^{-9}(C^o)^{-1}$. There are several background articles that provide further description of this specific technique, apparatus used and actual thermal expansion measurement results for different materials (6-8). Bethold et.al, have reported a daily length change of $(5.64 \pm 0.03) \times 10^{-9}$ for an Invar LR-35 sample, which was triple heat treated by the water quench method previously discussed. This is roughly equivalent to a 2 PPM growth per year.

Boeing conducted thermal expansion tests on JPL provided WF/PC "flight" parts. An example of one of these parts is shown in Figure 2. Other examples of flight parts are shown in Figure 3. A Boeing developed 19 - channel laser interferometer was used. This 19 - channel measuring device allows the simultaneous measurement of multiple displacement paths during each test. General Dynamics-Convair has used a Hewlett Packard single channel interferometer on JPL samples.

Test Description

Although there were slight variations in test directions for some samples, the following list of tasks is a pretty good description of what was done with most samples:

1. Measured the thermal coefficient of expansion of the specimens over the temperature interval of -73C to 66C.

2. The specimens were initially brought into thermal equilibrium at 20C and the reference length ($\Delta L/L$=00) established.

3. Specimen $\Delta L/L$ was measured at the following equilibrium temperatures in the sequence here given: 20C, 35C, 50C, 66C, 20C, 5C, -10C, -25C, -40C, -55C, -73C, 20C.

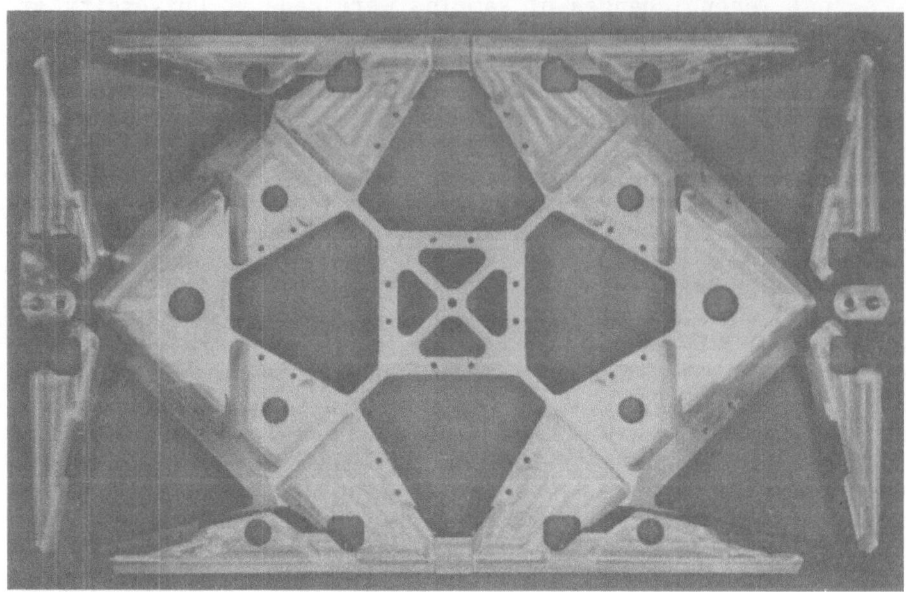

Fig. 2. WF/PC low-C Invar optical bench bulkhead. Outside dimensions are approximately 15" x 24". Three orthogonal planes of machining were necessary.

Fig. 3. Typical WF/PC Invar parts for mirror mounting and clamp assemblies. Large ring (at left) outside diameter is approximately 6 inches.

Some specimens were cycled 20 times through the following temperature sequence: 20C, 66C, -73C, 20C, ΔL/L values were measured with the specimens at thermal equilibrium at the end of the 1st, 5th, 10th and 20th cycle, at 20C.

The rate of temperature change was prescribed to be less than 2C/minute when proceeding between equilibrium temperatures or during thermal cycling. The end point temperatures (-73C & 66C) were to be held for a minimum of 10 minutes during thermal cycling.

The tolerance on achieving the required equilibrium and cycling end point temperatures was ±2C and thermal equilibrium was defined to be at the end of an interval of at least 20 minutes during which the sample temperature did not change by more than 1°C.

The following measurement techniques and instrumentation applied to the above mentioned tasks.

1. The test samples were provided by JPL. The sample sizes were specified by contractor. JPL performed heat treat cycles on all samples. Variations in heat treat procedures were performed but not identified to contractors.

2. The expansion parameter measured was a sample change in length. The resultant thermal strain (ΔL/initial length @ 20C) was to be measured to a resolution better than 0.1 PPM. The resolution in the CTE would, therefore, be on the order of 0.01 PPM/C or better.

3. Each sample was instrumented with thermocouples and/or thermistors. These thermocouples/thermistors provided sample temperature data.

4. All data were submitted in graphical and tabular form. The graphical data were a plot of the thermal strain as a function of temperature. The tabulated data included the measured strains, calculated to 0.01 PPM or better, and the sample temperature, measured to 0.1C.

RESULTS AND CONCLUSIONS

A compiled listing of thermal expansion results obtained for the eighteen specimens is shown in Table 2. Two examples of thermal expansion results in graphical form are presented in

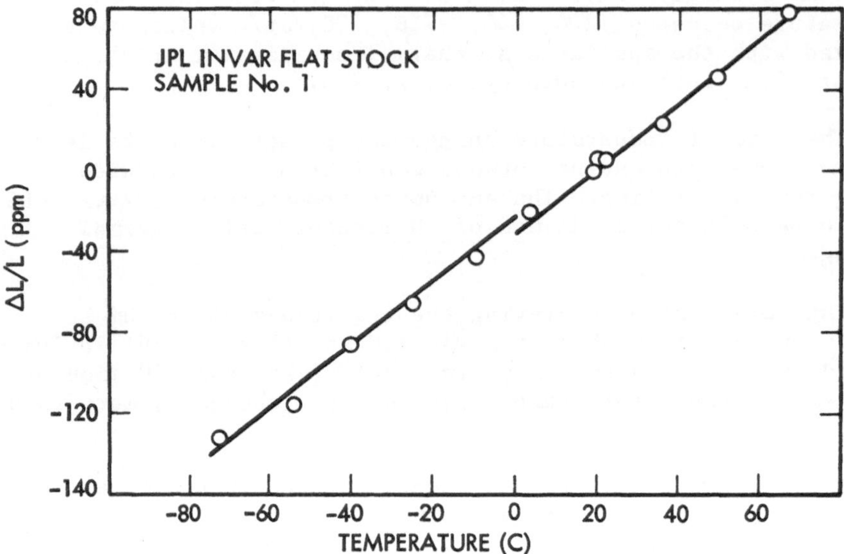

Fig. 4. Low-C Invar rod stock thermal expansion.

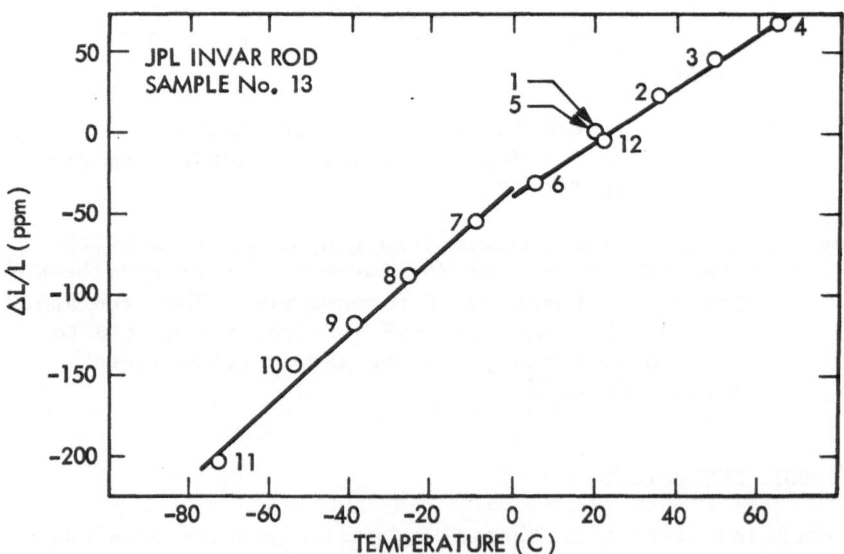

Fig. 5. LR-35 Invar flat stock thermal expansion.

Table 2. Coefficient of Thermal Expansion Results

JPL ID NO.	MATERIAL (ORIENTATION)*	TREATMENT CYCLE**	MEASUREMENT TECHNIQUE	MEAN THERMAL STRAIN ($PPM/^{o}C$)			
				$-70^{o}C \rightarrow 0$		$0 \rightarrow +70^{o}C$	Alternate T^{o} Range
1	Low C Invar (L)	OIL QUENCH	LASER OPTICAL COMPARATOR(COI)	1.56		1.61	
2	Low C Invar (L)	OIL QUENCH	LASER OPTICAL COMPARATOR (COI)	1.58		1.60	
3	Low C Invar (L)	OIL QUENCH	FABRY-PEROT ETALON (U OF A)	1.64		1.49	
4	Low C Invar (L)	ANNEAL	FABRY-PEROT ETALON (U OF A)		3.0		$(20 \rightarrow 40^{o}C)$
5	Low C Invar (LT)	ANNEAL	DILATOMETER (SIMONDS STEEL)		-0.63		$(10 \rightarrow 30^{o}C)$
6	Low C Invar (L)	ANNEAL	QUARTZ DILATOMETER (BOEING)		~1.2		$(-100 \rightarrow +50^{o}C)$***
7	Low C Invar (LT)	ANNEAL	QUARTZ DILATOMETER (BOEING)		~1.5		$(-100 \rightarrow +50^{o}C)$***
8	Low C Invar (ST)	ANNEAL	QUARTZ DILATOMETER (BOEING)		~0.95		$(-100 \rightarrow +50^{o}C)$***
9	Low C Invar (LT)	ANNEAL	LASER INTERFEROMETER (BOEING)		1.46		$(5 \rightarrow 20^{o}C)$
10	Low C Invar (LT)	ANNEAL	LASER INTERFEROMETER (BOEING)		1.46		$(5 \rightarrow 20^{o}C)$
11	Low C Invar (LT)	ANNEAL	LASER INTERFEROMETER (BOEING)		1.49		$(5 \rightarrow 20^{o}C)$
12	Low C Invar (LT)	ANNEAL	LASER INTERFEROMETER (BOEING)		1.49		$(5 \rightarrow 20^{o}C)$
13	LR-35 (L)	OIL QUENCH	LASER OPTICAL COMPARATOR (COI)	2.21		1.68	
14	LR-35 (L)	OIL QUENCH	LASER OPTICAL COMPARATOR (COI)	2.16		1.63	
15	LR-35 (-)	WATER QUENCH	LASER INTERFEROMETER (CONVAIR)				
16	LR-35 (-)	WATER QUENCH	LASER INTERFEROMETER (CONVAIR)				
17	INVAR 36-free machining (-)	ANNEAL	LASER OPTICAL COMPARATOR (COI)	2.29		2.29 2.29	
18	INVAR 36-free machining (-)	ANNEAL	LASER OPTICAL COMPARATOR (COI)	2.50		2.29	

* ORIENTATION, L-LONGITUDINAL (IN GRAIN OR ROLLING DIRECTION), LT-LONGITUDINAL TRANSVERSE, ST- SHORT TRANSVERSE AND (-) UNKNOWN

** SEE TEXT FOR COMPLETE DESCRIPTION

*** BOEING PRELIMINARY DATA

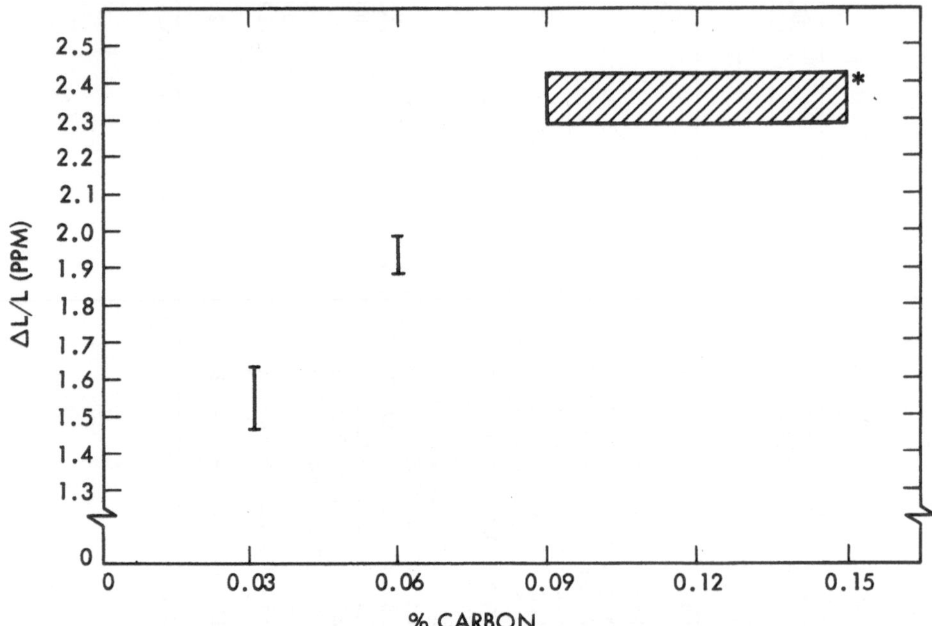

Fig. 6. Semi-quantitative graph showing the relationship between Invar carbon content (%) and thermal expansion over -70°C to +70°C. Data are from oil-quenched and/or annealed samples only. *Possible carbon content range is shown.

Figures 4 and 5. Based on these results and other test observations, the following results/conclusions were drawn:

1. The subject alloys did not undergo a phase change (at least not above $-70^{o}C$) so that CTE values remained consistent over as many as twenty thermal cycles. No conclusive evidence of any permanent dimension change (ΔL) after thermal cycling was apparent.

2. "Relative" length change measurements were in good agreement with results from "actual" length measurement technique results. Considerable cost savings were realized in utilizing "relative" over "actual" measurement techniques.

3. Significant problems occurred at University of Arizona in their attempts to obtain and maintain good optical contact to the invar specimens with mirror endplates. Only realistic appearing results from two of several provided samples have been included.

4. Results of quartz dilatometric measurements are highly suspect. Agreement with other measurement technique results was poor.

5. Lowest coefficient of thermal expansion was obtained for the Low-C annealed invar. Oil quench operations appeared not to alter (lower) the CTE to any degree.

6. General confirmation of the impact of carbon content on thermal expansion was confirmed, i.e., the lower the carbon content, the lower the CTE as shown in Figure 6. In the carbon content range studied the CTE sensitivity was about 1 PPM increase for a corresponding 0.1% increase in carbon.

ACKNOWLEDGEMENT

Research described was carried out by the Jet Propulsion Laboratory/California Institute of Technology under a contract with National Aeronautics and Space Administration.

References

1. D. D. Smith, "A Space Stable Optical Bench Structure For the Space Telescope Wide Field and Planetary Camera," Sampe Symposium, (1980)

2. B. S. Lement, B. L. Averbach, M. Cohen, "The Dimensional Behavior of Invar," Transactions of the A.S.M. Vol. 43, 1951, Pages 1072-1097.

3. J. W. Berthold III, S. F. Jacobs, and M. A. Norton, "Dimensional Stability of Fused Silica, Invar, and Several Ultralow Thermal Expansion Materials," Metrologia 13, 9-16 (1977).

4. H. Masumoto, "On the Thermal Expansion of the Alloys of Iron Nickel and Cobalt, and the Cause of the Small Expansibility of Alloys of the Invar Type," Science Reports, Tohoku Inperial University, V.20, P. 101, (1932).

5. E. G. Wolff, C. S. Susskind, D. L. Dull and S. A. Eselun, "Processing Effects on the Dimensional Stability of Invar," Second International Conference on Mechanical Behavior of Materials, Boston, MA, 16-20 August 1976.

6. W. S. McCain et at., "Mechanical and Physical Properties of Invar and Invar-type Alloys." Battelle Memorial Institute Report AD 474 255, August 1965.

7. S. F. Jacobs, J. N. Bradford, and J. W. Berthold III, "Ultraprecise Measurement of Thermal Coefficients of Expansion," Applied Optics, Vol., 9, Page 2477, November 1970.

8. J. W. Berthold III and S. F. Jacobs, "Ultraprecise Thermal Expansion Measurements of Seven Low Expansion Materials," Applied Optics, Vol. 15, No. 10, October 1976.

9. S. F. Jacobs, "Measurements of Ultrasmall Displacements," Optical Engineerings, Vol. 17, No. 5, Sept.-Oct. 1978.

DEVELOPMENT AND TESTING OF ZERO CTE MATERIALS

E. G. Wolff

Aerospace Corporation
2350 El Segundo Blvd.
El Segundo, California 90245

INTRODUCTION

Structural materials with reduced coefficients of thermal expansion (CTE) represent improved performance for many industrial and aerospace systems. For example, the discovery of Invar by Guillaume in 1896 facilitated the development of optical metering and support structures, laser housing, geodetic tapes, microwave components and many instrument parts[1]. Low expansion ceramics such as cordierite are desirable substrates for automotive catalytic converters[2], while hafnia-titania compounds are of interest as flame tube liners[3], again because of the thermal shock-environment. Low expansion composites are needed in many aerospace applications[4]. Mirror substrates which will operate at non-ambient temperatures are primary candidates for zero CTE materials[5].

In general, the overall requirement is dimensional stability when one or more of the following change: temperature, time, chemical and radiation environment (e.g. moisture) and stress or pressure, both internal and external. All materials respond dimensionally to any of these parameters, sometimes irreversibly (e.g. microcracking). Consequently, both the development and testing of zero CTE materials is complex.

Instantaneous zero CTE's already exist, e.g. in SiO_2, when the slope of $\Delta L/L$ versus T curve reverses sign. The practical objective is to develop a volumetric zero CTE (implying isotropy) over a broad temperature range. Such a material does not yet exist, but alloys based on Invar, glass/ceramics and some types of composite materials and structures appear promising.

THEORY

Figure 1 illustrates a zero CTE prediction in the form of a mixture of ULE and 13 vol % Fe + 36 Ni[6]. In practice, zero CTE development is subject to uncertainties in various theoretical models[7] and auxiliary effects which become more pronounced as zero CTE approaches. Most models are limited by assuming elastic, isotropic and well characterized and bonded constituents. For example, the cured shapes of unsymmetric laminates do not readily conform to the predictions of classical laminate theory[8]. Parameters which can affect the CTE include stoichiometry[6], phase transformations[7,9], microstructure[10], impurities[11,12] microcracking[13,14], stress[15], thermomechanical treatments[1], grain size, porosity, and self weight effects. With composites there are additionally plastic flow[16], viscoelasticity[17], anisotropic constituents[18], debonding[19], Poisson effects and relief of residual stresses. Edge and end effects alter the residual stress state of a test sample[20], so that there may be free edge delaminations[21], surface distortions, sites for preferential microcracking and hence stiffness and CTE variations. Thermal flexure of asymmetrical specimens can also induce significant test errors[22].

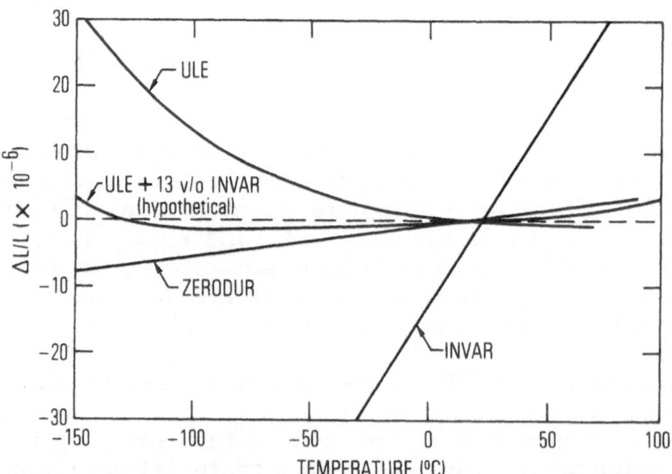

Fig. 1 Thermal expansion of common low CTE materials and a hypothetical mixture to give zero CTE (25 to -130°C)[6]

ALLOYS

Invar type alloys (nominally Fe + 36Ni) have the lowest CTE values of metallic systems. Invar sheet decarburized to ≅ 0.01 wt % C, and treated to 845°C, water quenched, heated to 95°C, and air cooled, showed a CTE of 0.01×10^{-6} °C^{-1} (5-30°C)[11]. Further decarburization and optimized thermo-mechanical treatments could conceivably lower this value, or perhaps stretch the applicable temperature range. However, treatments which induce low CTE's (e.g. water quenching from the γ range) may be incompatible with temporal stability, which usually requires slow cooling[23].

Super Invar (nominally Fe-30Ni-6 Co) was much studied in the 1930's[24-26], and has been identified (along with Zerodur, a Schott glass ceramic) as the material with best temporal stability at room temperature[27]. It exhibits both low positive[24] and negative[12] CTE values. The $\Delta L/L$ vs T curves are discontinuous below 500°C, first because of the paramagnetic to ferromagnetic (A_2) inflection temperature in the region 300-500°C. The $\gamma + \alpha \rightarrow \alpha$ transition (Ar_3) generally occurs below ambient, but can start as high as +100°C[25]. The CTE and the Ar_3 start temperature are sensitive to plastic deformation, heat treatment, alloying additions and trace impurities. C, Cu, Mn, Mo, Nb and Re tend to lower the Ar_3 but raise the CTE; Co and Cr both lower the CTE but only Cr lowers the Ar_3. Si and Ti tend to raise the Ar_3. Chromium, to make "Stainless Invar" can give negative CTE's with no Ar_3 problem above -150°C[9,26]. For example, alloy #51 with 37Fe, 54 Co and 9 Cr showed a CTE of -1.2×10^{-6} °C^{-1} near 20°C[26]. The objective of a simultaneous Ar_3 below -100°C and zero CTE for an annealed alloy over a temperature range of at least 60°C appears to be possible within the Fe-Ni-Co-Cr system.

Studies in our laboratory directed to a low CTE metal matrix for composites have involved hot pressed, atomized, pre-alloyed Fe-Ni-Co powders. Problems with both homogenization and chemical analysis were found. Spectroscopic methods were most accurate for impurity levels and atomic absorption methods were best for alloying elements, but there were discrepancies with intermediates such as Mn and Si. When hot pressed at 880°C, 1 hr and 4900 psi in a graphite die in vacuum, the density was 6.69 g/cc (~ 94% of theoretical). The same powder was hot pressed at 1200°C and 3000 psi in argon at MIT and also tested with a double Michelson interferometer[19]. Initial CTE's were $< 10^{-6}$ °C^{-1}, but after a cool to -100°C the CTE's rose to the $3 - 4 \times 10^{-6}$ °C^{-1} range over ± 100°C (Figure 2). The Ar_3 on-set temperature is clearly above -100°C for this composition, and the extent of the transformation would depend on the cooling/heating rates.

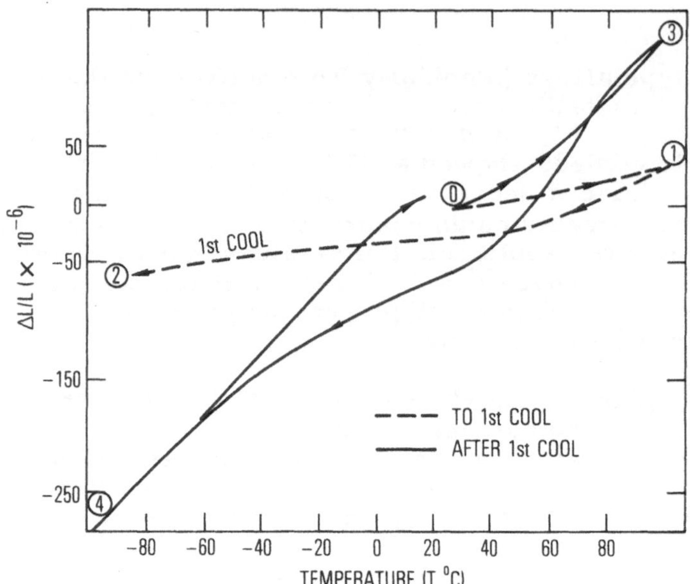

Fig. 2 Semi-quantitative summary of Thermal Cycling Behavior of a Hot Pressed Super Invar Powder (28-33, Ni, 6.1-11 Co, 0.06-0.28Cr, 0.006 Mn, 0.01-0.25 Si, 0.01 C, 0.027Mo, 0.013 Ti, 0.055 Cu, 0.013 O, 0.024-0.23 Al, Bal Fe

CERAMICS AND GLASSES

Due to the needs of mirror substrates, more work towards zero CTE has been carried out in the fields of glasses and ceramics than any other materials systems. Fused silica, Zerodur, ULE (Corning SiO_2 + 7% TiO_2), and CER-VIT (Owens-Illinois glass-ceramic) are notable examples. Extrusion helps to orientate cordierite crystals (parallel "c") to give particularly low values of CTE[2]. The latter is sensitive to alkali with K^+ promoting a negative CTE at low temperatures, and Cs^+ a high value. Other laboratories are working to improve cordierite by not only getting a CTE nearer zero but also, through substitution of chemical modifiers, by reducing internal stresses to avoid microcracking[28]. Other ceramics with low positive CTE's of commercial interest include silicon nitride, which can be made to 100% density in an amorphous state by a CVD process to give high strength and hardness as well as a CTE of $\sim 10^{-6}\ {}^oC^{-1}$ [29]. Ceramics of interest for negative CTE's include β-eucryptite (Li_2O-Al_2O_3-2 SiO_2)[30], β-spodumene (Li_2O-Al_2O_3-4 SiO_2)[6] and their mixtures[31], La_2O_3,[19,32] Hf-Ti-O[3], Ta-V-O[19], Ta-W-O and Hf-W-Ta-O compounds[33]. Microcracking has been used to explain many negative CTE data in ceramics[6,19]. There are possibilities of predicting the expansions due to microcracking in graphite reinforced plastics[34] but with ceramics, there is also the variable of grain size. It determines not only onset temperature of microcracking but also the extent, since hot pressed samples show different $\Delta L/L$ vs. T curves than cold pressed and sintered ones. Grain boundary impurities may also cause microcracking, either through thermal expansion mismatches or through promotion of localized densification[35]. Mirror uniformity has required extensive attention to chemical homogeneity and methods of nondestructively measuring CTE variations[36,37].

FIBER REINFORCED PLASTICS

The low CTE values obtainable in two directions with graphite reinforced plastics are well known[4]. The CTE can be readily predicted by computer programs which derive complete laminate constitutive relations from basic lamina properties[34,38]. To predict the onset of microcracking, and hence a deviation from the predicted CTE, one calculates the temperature change required to effect a transverse stress equal to the ultimate strength of the matrix, e.g. 4000 psi for epoxy. For example, an HMS fiber (60 vol %) in a 3501-6 epoxy matrix with $[90/\pm 45/0]_s$ layup will microcrack below -44°C if it were a symmetric plate. However, a 4-ply system, as might be used in a thin walled tube, is unsymmetric and cracking will start sooner; at -25°C. We have further developed finite element analyses to account for the constrained edge effects of the tube ends; there the microcracking could start as high as -12°C. Achievement of unidirectional zero CTE is possible by varying θ in a $(\pm\theta)_s$ fiber orientation (Figure 3). A $[0, \pm\theta, 0]$ layup may be preferable in practice because of the low value of Δ CTE/$\Delta\theta$ which relieves fabrication tolerances on θ. Close control of V_f, E_f, fiber CTE and porosity values are also needed.

GRAPHITE FIBER REINFORCED METALS

Graphite-magnesium is the first fiber reinforced metal matrix composite system that offers a realistic possibility of zero thermal expansion coefficient (CTE) in the fiber direction[39]. Elasto-plastic effects during thermal cycling are of particular interest because they are not superimposed on a large expansion behavior. The low stiffness and strength of magnesium (relative to other structural metals) mean that plastic flow, microcreep and thermal fatigue effects are likely to show up at low temperatures (e.g. < 100°C). Figures 4 and 5 give $\Delta L/L$ in the fiber direction vs T during two thermal cycles for a unidirectional pitch graphite reinforced magnesium alloy. ULE mirrors were mounted at the ends of a 12" long rod for use with reflected beams from a laser interferometer system. It is seen that the

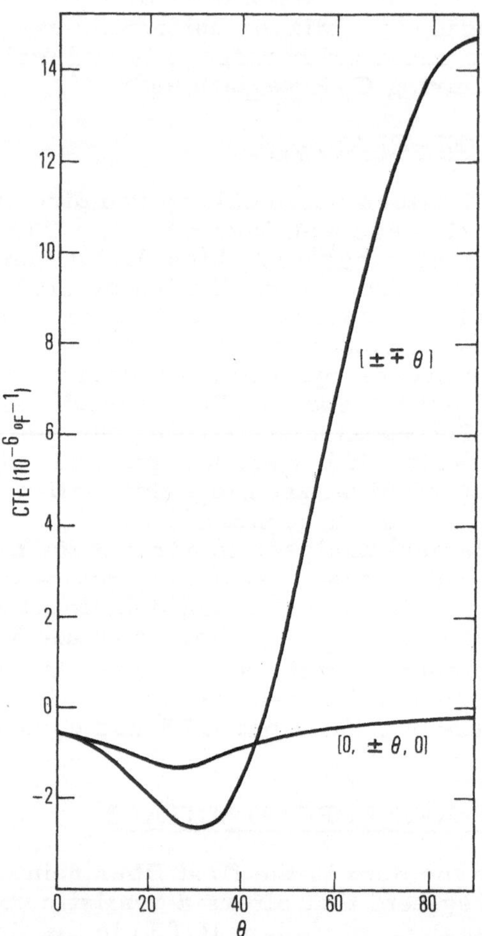

Fig. 3 Variation of CTE in the θ° direction with ply orientations

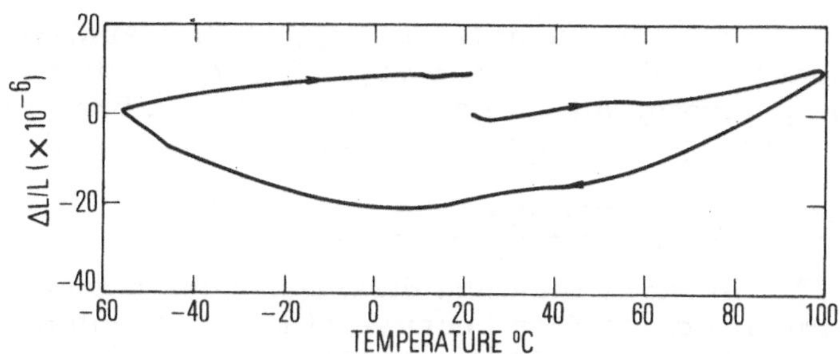

Fig. 4 Thermal cycling of a pultruded VS 0054 (Pitch 100), (40 vol %), reinforced AZ91C Mg Alloy rod in the fiber direction with initial heating. Rates ~ 1.2°C/min

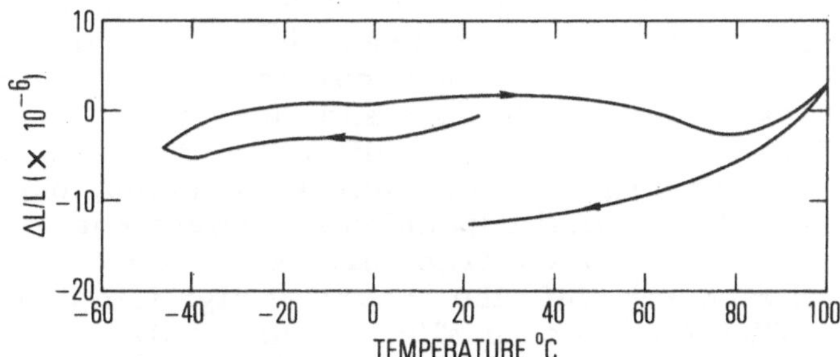

Fig. 5 Thermal cycling of the Gr/Mg rod with initial cooling following the cycle of Figure 4

initial heating produces the low CTE predicted by the Turner equation ($\sim 0.9 \times 10^{-6}\ {}^{o}C^{-1}$), suggesting elastic deformation. Sufficient cooling, however, produces a negative CTE, indicative of plastic flow of the matrix and a dominance of the negative CTE of the fiber. The results of these and more extensive thermal cycling studies will be reported elsewhere. They are explainable in terms of current theories of elastic-plastic flow in composites. Changes in microstresses caused by the heat treatments can be treated by rigid body translations of generalized yield surfaces in the direction of the hydrostatic axis[16]. Other analytical efforts account for the fiber anisotropy[18]. Weak fiber-matrix interfaces may become an analytical as well as a practical problem; usually a perfect interface has been assumed.

GRAPHITE-CARBON COMPOSITES

Zero CTE in more than one direction does not seem likely in graphite-metal composites because of the high volume loadings of graphite fibers required in the longitudinal case. With graphite reinforced carbon, we already have multi-directional fiber orientations filled with various carbon binders (and often considerable porosity). Reported work in this area is limited[13,40]. The general shape of the $\Delta L/L$ vs T curve is similar for all pure C/C composites with CTE's falling between the "a" and "c" direction values of graphite. Studies in our laboratory on felted and 5D-Gr/C materials suggest that negative CTE values are common below ~300°C (e.g. -0.1 to -0.7 x 10^{-6} °C^{-1}). CTE values can also be as high as ~0.5 x 10^{-6} °C near 20°C[40]. A feature of most C/C composites as a consequence of current fabrication methods is porosity. Cracks are likely to open and close during thermal cycling[13]. Free end effects are significant with C/C composites due to the relatively large microstructural features, e.g. bundles of fibers. Accurate thermal expansion measurements may require corrections of over 7% for the free surface involved[14]. Creep due to residual stress relief also occurs at high temperatures[41]. Isotropic zero CTE Gr/C composites would require some ingenuity, e.g. metallic coatings on the negative CTE graphite, coatings on the final composites, or modification of matrices and infiltration procedures. Expense, microscopic homogeneity and reproducibility are continuing problems, but tailoring to zero CTE in several directions is a distinct possibility. Boron nitride may be a future complement to graphite fiber systems since the CTE in the "a" direction of the hexagonal form is also negative.

LARGE STRUCTURES

Adhesively bonding the ends of two graphite epoxy tubes with a plug produces a near zero CTE structural element[28]. One tube has a positive CTE, the other a negative one, and the ends are cut off to make a net zero linear CTE. The idea of a slightly negative CTE graphite epoxy tube with positive CTE end fittings, e.g. of Al, Ti, or molded graphite cloth is a common one for large space structure design concepts[42-44]. Glass cloth is often wrapped around the tube to minimize the thermal stresses due to the CTE mismatch. Another concept is to use glass and graphite fibers together in a plastic matrix to arrive at a (linear) zero CTE. Such a composite can be made inexpensively by pultrusion methods for truss structures, antenna mounts, and optics supports[45]. For a zero CTE, the composition should be 14.2 (vol) % glass, 38.5% graphite and 47.3% resin. Uniform and parallel fiber distribution are essential. The allowable radius of curvature (e.g. in stored, furlable antenna ribs or end fittings) will be reduced as the percentage of graphite increases.

METAL CERAMIC SYSTEMS

When a negative CTE ceramic is embedded in a continuous metal matrix, the prospect for isotropic zero CTE with high thermal conductivity is most promising. Except for certain Fe-Ni-Co-Cr alloys, a negative CTE metallic phase appears unlikely. There has been development of related fabrication methods, but an optimum composition has yet to be chosen. In a study aimed at mirror applications[46], 2 and 4 wt % CuO were added to Li-Al-Si and Zn-Al-Si based glasses and melted, and cast or drawn. The latter crystallized (at 625-725°C) to give very low CTE values (e.g. $-0.14 \times 10^{-6}\ {}^{\circ}C^{-1}$ for 0 - 300°C). Heating was used to generate CuO or Cu films, (depending on the atmosphere) on the surface via diffusion. Glass inhomogeneity and film discontinuities were two problems encountered, but the self generated metallic surface showed generally excellent fiber-substrate adhesion. Other systems studied include Cu, Ni and steels with glass[47,48], Cu-TiO_2[49], and Nb_2O_5 in Ni and Super Invar[19]. Dissolution of the metal in the glass or ceramic[50], particle-void interactions[51], yielding of the metallic phase, and fracture of the ceramic[19] are potential problems.

Precoated (with Ni) Nb_2O_5 powders and Nb_2O_5/Super Invar powder mixtures were hot pressed at MIT[19] and tested in our laboratory for CTE. The end face displacements were monitored in a double Michelson interferometer during thermal cycling in the range $\pm 100^{\circ}C$[53]. The results in Figure 6 appear to fit a simple (modulus independent) rule of mixtures or upper and lower bounds of the CTE of an isotropic composite of arbitrary microstructure[7]. The upper bound is identical to the Kerner equation[10] for spheres of material A surrounded by shells of material B. (Values of E_{metal} = 30 msi and $E_{Nb_2O_5}$ = 23.5 msi were assumed[54].) The Super Invar behavior was identical to that shown in Figure 2. Still better agreement with predictions might be possible by considering that the small amount of porosity reduces the effective values of constituent volume fractions. The CTE and moduli are both temperature dependent, and some microcracking within the polycrystalline Nb_2O_5 particles is possible due to its inherent crystal anisotropy[6,19]. Figure 7 illustrates the apparently negative CTE below about -40°C found only on first cooling below ambient for a typically high Nb_2O_5 content. Evidence of microcracking via an opto-acoustic technique[34] was not confirmed. Yielding of the metal should have occurred in subsequent cycles. Lack of hysteresis in subsequent cycles also reduces the possibility of debonding. Nb_2O_5 sample anisotropy was not introduced as a result of hot pressing. Further development will utilize fine grained or single crystal ceramic dispersions with negative volumetric CTE's. A tradeoff between debonding due to a stiff metallic matrix and plastic flow with a low E matrix may be necessary.

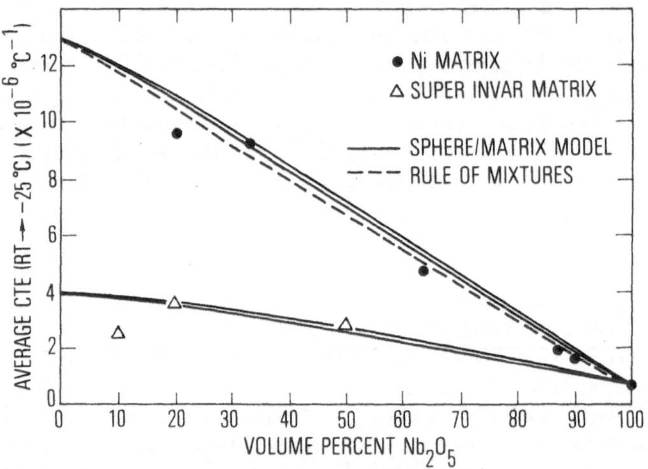

Fig. 6 Average CTE (0 ± 25°C) of Ni-Nb_2O_5 and Super-Invar-Nb_2O_5 composites as a function of volume fraction Nb_2O_5

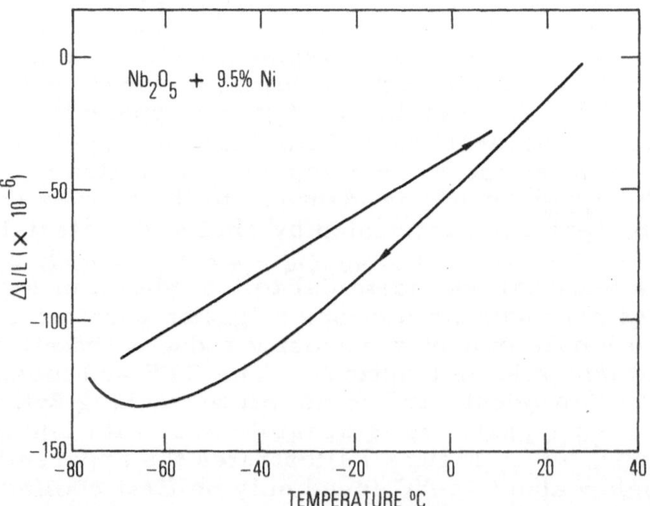

Fig. 7 ΔL/L vs T for Initial Cooling below 25°C for a hot pressed Ni-Nb_2O_5 mixture

TESTING

Sensitivity in Δl measurements need not be a problem for zero CTE studies. Gravity research, for example, is helping to develop laser interferometer techniques which can detect Δ l's of $<10^{-15}$m[55]. For CTE, however, the temperature measurement is important and also temperature induced errors in an interferometer system, such as misalignments, system drifts and sample support motions. A laser also produces heating effects on an Angstrom level (Figure 8). The initial displacement is linear with time and directly proportional to the CTE of the substrate[39,53]. On the other hand, if the laser beam is too weak, signal to noise ratios drop and optical decouplers cannot be used. Although polished end faces are best for laser reflection, cold working may induce residual stresses which affect the CTE[23]. ULE mirror attachments should be spring loaded since adhesives (e.g. Varian "TORR-Seal") introduce small errors due to their high CTE. Motions of the sample due to heating necessitate lenses to focus the interferometer beams to give the "cats eye" parallel back reflection. There is no lens material with a zero "optical path length change per unit temperature change" so that optical systems symmetry is required. Retro-reflectors should not be used in zero CTE work due to the difficulty of eliminating thermal and hence index of refraction gradients. X-ray diffraction methods are highly sensitive to lattice parameter changes with temperature but also, unfortunately, to local composition and crystallite orientation variations.

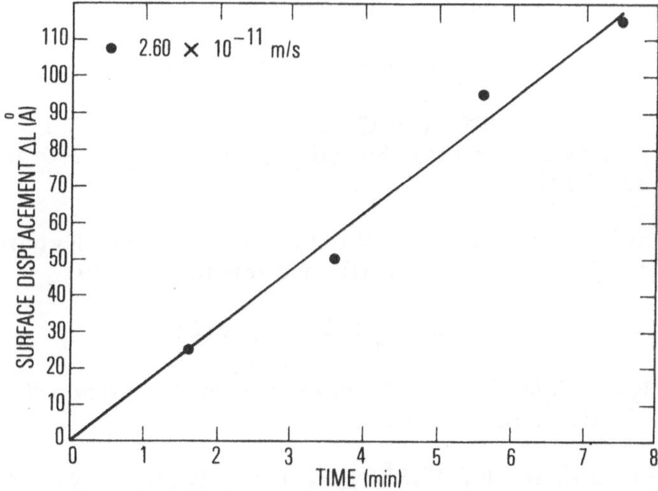

Fig. 8 Surface swelling of a 500Å Al coated SiO_2 rod end when illuminated by 0.3mW He-Ne laser beam

CONCLUDING REMARKS

Fused silica is likely to remain the primary standard (NBS-SRM-739) for low expansion materials for some time. ULE glass (Corning SiO_2 + 7% TiO_2) has been suggested for lower CTE values (~3×10^{-8} $^{o}C^{-1}$) near 25°C, but it expands substantially at lower temperatures (Figure 1). There is a chance for any of the various classes of material discussed in this paper to provide a replacement from the fused silica as standard. Homogeneity and reproducibility will require controllable chemical compositions, simple thermal-mechanical processing and long term thermal cycling stability. Ultimately, its CTE measurements should be tied to the laser wavelength standard of length.

ACKNOWLEDGEMENTS

This work was sponsored by the U.S. Air Force and the Defense Research Projects Agency under Contract F04701-80C-0081.

REFERENCES

1. C. W. Marschall and R. E. Maringer, "Dimensional Instability"Pergamon Press, p 268 (1977)

2. I. M. Lachman, R. D. Bayley and R. M. Lewis, Bull. Am. Ceram. Soc. 60 202 (1981)

3. S. R. Skaggs, Los Alamos Rept. LA-UR-77-1307 (1977) also Int. Coll. High Temp. Energy Sources, Odeillo, France June 1977

4. E. G. Wolff, Metal Progress 115 54 (1979)

5. F. Ayer, E. G. Wolff and G. C. Comisar, p 27 in "Thermal Expansion 6" Ed. I. D. Peggs, Plenum Press (1978)

6. G. F. Hawkins and E. G. Wolff, 7th Thermal Expansion Symposium, Chicago, Ill, November (1979)

7. D. K. Hale, J. Mat. Sci. 11 2105 (1976)

8. M. W. Hyer, Virginia Polytechnic Institute Report VPI-E-81-4 March 1981

9. P. Hidnert and R. K. Kirby, J. Res. NBS, Paper 2602 55 29 (1955)

10. M. DeAlmeida, R. J. Brook and T. G. Carruthers, J. Mat. Sci. 14 2191 (1979)

11. C. W. Marshall, SAMPE, 21 261 (1976)

12. A. I. Zakharov, A. M. Perepelkina and A. N. Shiryaeva, Metallovedenii i Termicheskaya Obrabtka Metallov, No. 6 62 (1974)

13. J. Jortner, Proc. Army Symposium Solid on Mechanics AMMRC MS-76-2, 81-97 (Sept. 1976)

14. J. Jortner and F. I. Clayton, JANAF Rocket Nozzle Matls. Rept. 1(5), (1979)

15. A. R. Rosenfield and B. L. Averback J. Appl. Phys. 27 51 (1979)

16. G. J. Dvorak and Y. A. Bahei-El-Din, J. Mech. Phys. Solids 27 51 (1979)

17. D. L. Flaggs and F. W. Crossman J. Comp. Mat. 15 22 (1981)

18. G. A. Gurtman, M. H. Rice and A. Maewal, Systems, Science & Software Inc., La Jolla, CA, Report SSS-R-81-4862, Feb (1981)

19. N. Saka and J. Boustani (MIT) private communication (1980) Also J. Boustani, M. E. (Eng) Thesis, MIT (1981)

20. A.S.D. Wang and F. W. Crossman, J. Comp. 2 300 (1977)

21. P. W. Hsu and C. T. Herakovich, J. Comp. 11 422 (1977)

22. H. M. Kural and A. M. Ellison, SAMPE 16(5) 20 (1980)

23. C. S. Susskind, D. L. Dull, S. A. Eselun and E. G. Wolff, Second Int'l Conf. on Mech. Behavior of Mat. (ASM) Boston, MA, Aug 16-20

24. H. Masumoto, 260th Science Report, Tohoku University, 20 101 (1931)

25. H. Scott, Trans AIME Inst. Metals Div. 89 506 (1930)

26. H. Masumoto, 323rd Science Report Tohoku University, 23 265 (1934)

27. J. W. Berthold, S. F. Jacobs and M. A. Norton, Applied Optics 15 1898 (1976)

28. B. Nelson, Perkin Elmer Corporation, private communication

29. G. Benson, ERG Inc., Oakland, CA, private communication

30. V. G. Christoserdov and V. I. Novgorodtseva, Catalysed Crystallization of Glass, Ed. E. A. Porai-Koshits, Trans. E. B. Uvarov, Consultants Bureau, New York p 154 (1964)

31. W. Ostertag, G. R. Fischer and J. P. Williams, J. Am. Ceram. Soc. 51 651 (1968)

32. S. N. Pandey and S. Singh, Bull. Am. Ceram. Soc. 58 184 (1979)

33. C. E. Holcombe, Bull. Am. Ceram. Soc. 59 1219 (1980)

34. S. A. Eselun, H. D. Neubert and E. G. Wolff, SAMPE 24 1299 (1979)

35. O. R. Biswas, S. Chandrateya and J. A. Pask, Bull. Am. Ceram. Soc. 58 792 (1979)

36. Papers on CTE Homogeneity at the 1971 and 1972 Spring Meeting, Optical Society of America

37. H. E. Hagy and W. D. Shirkey, Applied Optics 14(9) 2099(1975)

38. D. L. Reed, General Dynamics/FW #FZM-5494 (1970)

39. E.G. Wolff, E.G. Kendall and W. C. Riley, Proc. 3rd Int-Conf. on Comp. Mat., Paris 26-29 August 1980. Pergamon Press p 1140 (1980)

40. R. G. Naum, C. K. Jun and P. T. B. Schaffer in "Thermal Expansion-1971" Ed. M. G. Graham and H. E. Hagy, A.I.P., p 279 (1971)

41. J. Jortner, 15th Biennial Conf. on Carbon, U. of Paris, June(1981)

42. J. G. Bodle and L.M. Jenkins, 2nd AIAA Conference on Large Space Platforms, San Diego, CA paper AIAA-81-0445 Feb (1981)

43. B. Abt, K. Grunewald and M. Schneerman SAMPE 24 1333 (1979)

44. J. R. Lager 11th Nat. SAMPE Tech. Conf. ,SAMPE 11 471(1979)

45. T.J. Dunn, A.J. Cwiertney, V.L. Freeman and R. Johnson, NASA-CR-160558, (also NASA Tech. Briefs, p 323(1980)

46. P. P. Pirooz and G. Dube, AFML-TR-78-139 (1978)

47. B.R. Powell, G. E. Youngblood, D.P.H. Hasselman and L.D. Bentsen, J. Am. Ceram. Soc. 63 581 (1980)

48. L. N. Yagupol'skaya, S.V. Ivonova, E.V. Lysenko and N.I. Shcherban, Poroshkovaya Met. 11 pp 54-58 Nov 1976

49. K.G. Kumar, C. Pavithran and P.K. Rohatgi, J. Mat. Sci. 15 1588 (1980)

50. R. Z. Vlasyuk, E. S. Lugovskaya and I. D. Radomysel'sky, Poroshkovaya Met. 5 (101) 43 (1971)

51. I. W. Donald and G. Pollard, Powder Met. 18 (35) 32(1975)

52. V. N. Podgorkova and V. G. Mel'nikov, Poroshkovaya Met. 11 102 (1976)

53. E. G. Wolff and S. A. Eselun, SPIE Proceedings #192 (Interferometry) 204 (1979)

54. D. L. Douglass, J. Less Common Met. 5 151 (1963)

55. H. Boersch, H. J. Eichler, M. Pfundstein and W. Wiesermann, IEEE J. of Quantum Electronics QE-10 (6) pp 501-504 (1974)

THERMAL EXPANSION BEHAVIOR OF DENSE AND POROUS ALUMINUM POWDER COMPACTS

E. Ramasamy and P. Ramakrishnan

Department of Metallurgical Engineering
Indian Institute of Technology
Bombay 400 076, India

In the present investigation the thermal expansion behavior of massive aluminum after melting and of aluminum powder compacts of 5, 10, and 15% porosity was studied from room temperature to 873K using an automatic recording dilatometer in an atmosphere of nitrogen with a dew point of 228K. The influence of alloying elements on the dimensional changes during heating of porous aluminum powder compacts was investigated with reference to the additions of 4.4% copper powder for a binary system and 4.4% copper and 0.5% magnesium powder for a ternary system. The dimensional changes taking place during heating of the massive solid aluminum and of porous aluminum powder compacts with and without elemental additions are reported and discussed on the basis of thermal expansion of solid and porous aluminum, sintering, and alloy formation.

INTRODUCTION

The thermophysical behavior of a solid metal will depend upon its material characteristics, including the type of metal, whether it is massive or porous, the presence of other elements, their nature and quantity, various transformations taking place during heating, such as sintering, alloy formation, formation of liquid phase, and so on.[1] The magnitude of dimensional changes taking place during heating will also depend upon the numerous variables involved in the process, such as the initial characteristics of the powder from which the compacts are made, compacting pressure, porosity characteristics of the compacts, heating rate, temperature, time, and surrounding atmosphere. Thermal expansion of a solid metal is considerably different from that of a green compact made out of the powders of the same metal. Further small additions of other elemental powders for the purpose of alloying will alter the thermal expansion behavior remarkably. Sintering of metal powder compacts is an important processing step in the mass production of components by the powder metallurgy process. One of the convenient methods of studying sintering phenomena is dilatometry. In this method the dimensional changes during thermal treatment of the metal powder compact are used to investigate the sintering processes.[2,3] The thermal expansion behavior of a metal powder compact is different from that of a massive solid of the same material because of sintering in the powder compact.[4,5] In the present investigation the thermal expansion behavior of aluminum powder compacts of different porosities and also of compacts with additions of copper and magnesium powders was investigated and compared with that of massive solid aluminum.

EXPERIMENTAL

Atomized aluminum powder with particle size ranging from 150 μm to less than 45 μm used in this investigation is shown in Fig. 1. For the purpose of alloying, electrolytic copper powder of less than 38 μm and magnesium powder of less than 74 μm were used. Alloys of aluminum-4.4% copper and aluminum-4.4% copper-0.5% magnesium were made by mixing the elemental powders in a laboratory blender. Green compacts of 140 mm^2 cross section and 10.5 mm length were made using a hardened steel die and die wall lubrication. Compacts of various porosities were made by controlling the compacting pressure as well as the quantity of powder. The densities of the compacts were determined from the weights and dimensions of the compacts. Dense aluminum specimens were prepared by casting molten aluminum of the same composition into rods. The thermal expansion behavior of the solid and porous aluminum specimens was studied by observing the changes in linear dimensions using an automatic recording dilatom-

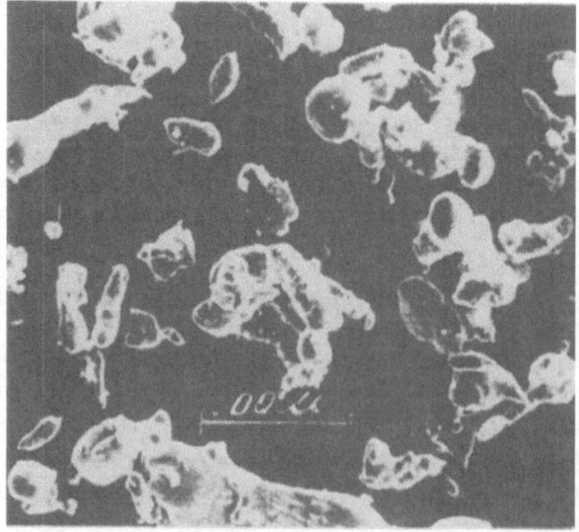

Fig. 1. Scanning electron micrograph of aluminum powder.

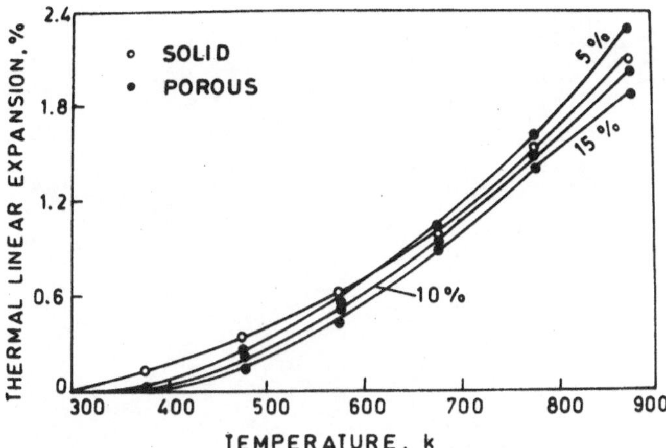

Fig. 2. Thermal linear expansion of solid and porous aluminum powder compacts of different porosities.

eter. The instrument was calibrated with pure aluminum solid samples. The specimen was heated at a rate of 22 degrees/minute in an atmosphere of nitrogen with a dew point of 228K. The $\Delta \ell / L_0$ was calculated for different temperatures from the continuously recorded linear changes and temperatures.

RESULTS AND DISCUSSION

The thermal expansion curves for solid aluminum and aluminum powder compacts of different porosities are shown in Fig. 2. The figure shows that both the massive solid aluminum and the porous compacts show expansion. The metal powder compact with 5% porosity shows more expansion than the solid material after a temperature of 673K, while the compacts with 10 and 15% porosity show less expansion than the solid material. The possible factors contributing to the thermal expansion of the specimens are thermal expansion of the metal and the expansion due to the gases entrapped during compaction. An increase in temperature increases the contributions of both of these factors. The larger expansion of the lower-porosity compacts at higher temperature is due to the additional expansion contribution due to the entrapped gases, which in this case are completely isolated. It has been reported in the literature for electrolytic copper powder that the pure expansion of the compact at the high compacting pressure far exceeds the value corresponding to thermal expansion at sintering temperature.[4] It is obvious that at higher compacting pressure the total porosity decreases, and many of the pores will be completely isolated. In the higher-porosity compacts, part of the thermal expansion of the

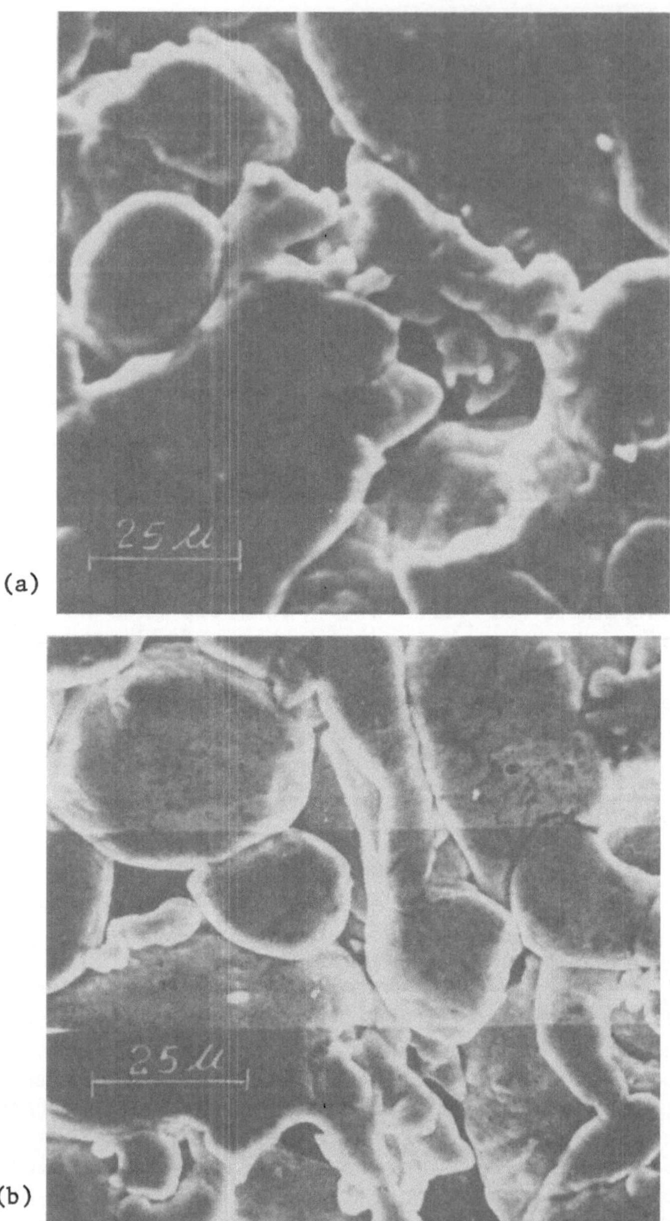

Fig. 3. Scanning electron micrographs of compacts of (a) 10% and (b) 15% porosity.

metal may be accommodated by the available porosity in the compact itself. Further, the contribution of the entrapped gases to the total expansion is less as a result of the interconnected porosity, which allows gases to escape. The larger amount of interconnected porosity in higher-porosity compacts is evident in the scanning electron micrograph of the 10 and 15% porosity compacts shown in Fig. 3. Hence, as the porosity of the compacts increases, the total thermal expansion decreases.

The thermal expansion curves for powder compacts of aluminum, aluminum-copper, and aluminum-copper-magnesium with 10% porosity are shown in Fig. 4. The Al-Cu system shows more expansion than the Al-Cu-Mg system, and both of them show more expansion than aluminum. Also, the Al-Cu-Mg system shows shrinkage at about 863K. The larger expansion of the powder compacts with elemental powder additions is mainly due to the additional expansion contribution due to the diffusion of alloying elements into aluminum. In the case of the Al-Cu system, the solid solution of copper in aluminum expands the aluminum particles. As the temperature is increased, aluminum-copper forms a eutectic at 821K and spreads into inherent voids in the compact, as a result of which new coarse pores appear at the position of the copper powder particles.[6] Again, the eutectic formation produces expansion as a result of volume increase upon melting of the Al_2Cu-Al eutectic, which pushes the particles apart. Recently various mechanisms that can produce swelling during the liquid-phase sintering of Al-Cu compacts as a result of eutectic liquid have been suggested, although it cannot yet be decided which of these mechanisms is mainly responsible for the swelling.[7] The lower expansion and shrinkage of Al-Cu-Mg powder compacts at about 863K is due to formation of a sufficient quantity of liquid phase, which results

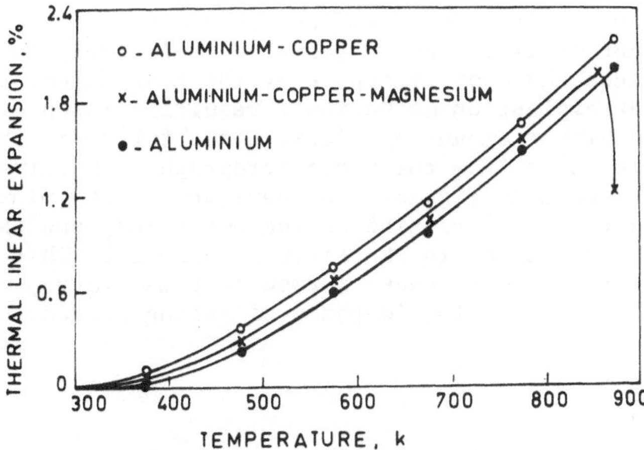

Fig. 4. Thermal linear expansion of powder compacts of aluminum, aluminum-copper, and aluminum-copper-magnesium alloy systems.

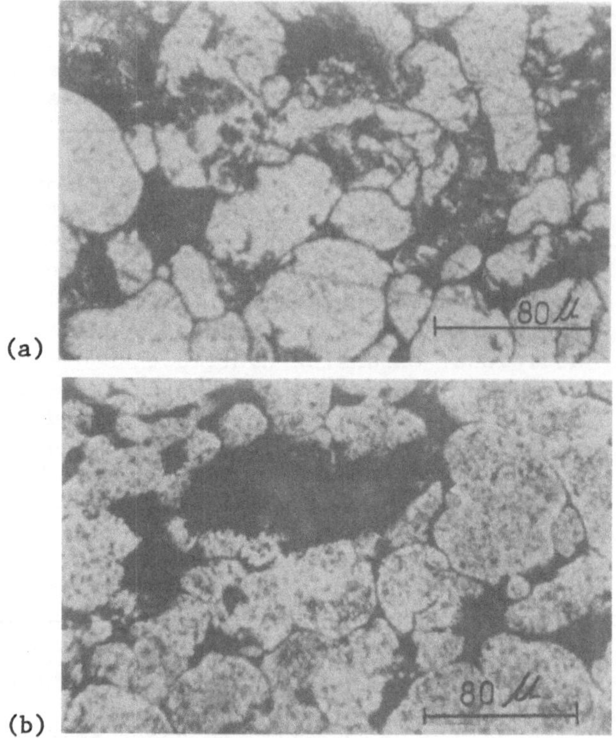

Fig. 5. Photomicrograph of specimens after heating to 873K. (a) Al-Cu, (b) Al-Cu-Mg.

in an overall shrinkage of the compact. Al-Mg binary eutectic forms at 723K, and the two ternary eutectics correspond to 723K and 781K. The larger amount of liquid phase formed at relatively lower temperature will enhance the mass transport by the rearrangement as well as solution reprecipitation mechanisms, resulting in an overall shrinkage. The higher amount of liquid phase formed in the Al-Cu-Mg system is also evident from the photomicrograph of specimens after heating to 873K shown in Fig. 5. The cavities in the microstructure are formed as a result of melting of the eutectics, and these pores are much higher and larger in the Al-Cu-Mg compact. This indicates the larger amount of liquid phase formed in this system, which promotes shrinkage by the liquid-phase sintering process.

CONCLUSIONS

Solid aluminum as well as well as porous aluminum compacts of different porosity expand on heating within the range of experimental investigation. The expansion of low-porosity aluminum is higher than that of solid aluminum, while the higher-porosity specimens show lesser expansion. Elemental additions to aluminum increase the expansion on heating as a result of solid state diffusion of copper to the aluminum lattice and Al_2Cu-Al eutectic formation. In the case of the Al-Cu-Mg system, the specimen expands initially and starts shrinking at about 863K as a result of liquid-phase sintering.

REFERENCES

1. Y.S. Touloukin, R.K. Kirby, R.E. Taylor, and P.D. Desai, "Thermophysical Properties of Matter - Thermal Expansion: Metallic Elements and Alloys," IFI/Plenum, New York, Vol. 12, p. 14a (1975).
2. Pol Duwez and Howard Materns, "A dilatometric study of sintering of metal powder compacts," Metals Trans., 185:571 (1949).
3. R. Sundaresan and P. Ramakrishnan, "Liquid phase sintering of aluminium base alloys," Intl. J. Powder Met. Powder Tech., 14:195 (1978).
4. H. Schreiner and R. Tusche, "Description of solid state sintering processes based on changes in length of the compacts made from different metal powders," Powder Met. Intl., 11: 52 (1979).
5. A.P. Sivitskii and N.N. Burtsev, "Compact growth in liquid phase sintering," Soviet Powder Met. and Met. Ceramics, 17:96 (1979).
6. T. Watanabe and M. Kanazawa, "Observation on the melt-off pores of sintered aluminium-5%copper alloy compacts made from mixed powder by the use of scanning electron microscope," Report of the Casting Research Laboratory, No. 24, p. 13 (1973).
7. W. Khel and H.F. Fischmeister, "Liquid phase sintering of Al-Cu compacts," Powder Met., 23:113 (1980).

EXPERIENCE WITH THE N.P.L. INTERFEROMETRIC DILATOMETER

Seton Bennett

National Physical Laboratory
Teddington
Middlesex
England

THE INTERFEROMETER

The NPL dilatometer measures linear thermal expansion interferometrically in the temperature range 20-500 $^{\circ}$C. This instrument, using a double-passed Michelson interferometer, achieves high accuracy and resolution while imposing a minimum of restrictions on the shapes and dimensions of specimens which can be investigated. Nearly one hundred samples have been measured, including blocks of various sizes, pieces of silica tubing and a section of the wall of a steel tank. The maximum cross-section which can be accommodated is 40 mm x 100 mm, while the minimum required is about 5 mm diameter.

The dilatometer has been described in detail previously (Bennett 1976). The specimen is placed on a base plate in a stainless steel vacuum oven (figure 1), and is heated by nichrome heaters wound on pyrophyllite bushes around the inner surfaces of a stainless steel enclosure. These heaters are connected in a series-parallel arrangement with a balancing rheostat to produce a uniform temperature within the oven. Specimen temperature is measured with a number of thermocouples which can be attached to the surface of the specimen or inserted into drilled holes. These sensors are regularly calibrated at NPL and are connected to a commercial ice-point reference unit.

The interferometer is constructed as a single, cemented unit (figure 2) consisting of three beam-splitters and a large cube-corner reflector. It measures the normal motion of the upper face of the specimen relative to the base plate, and is insensitive to small tilts of the reflecting surfaces. This insensitivity to

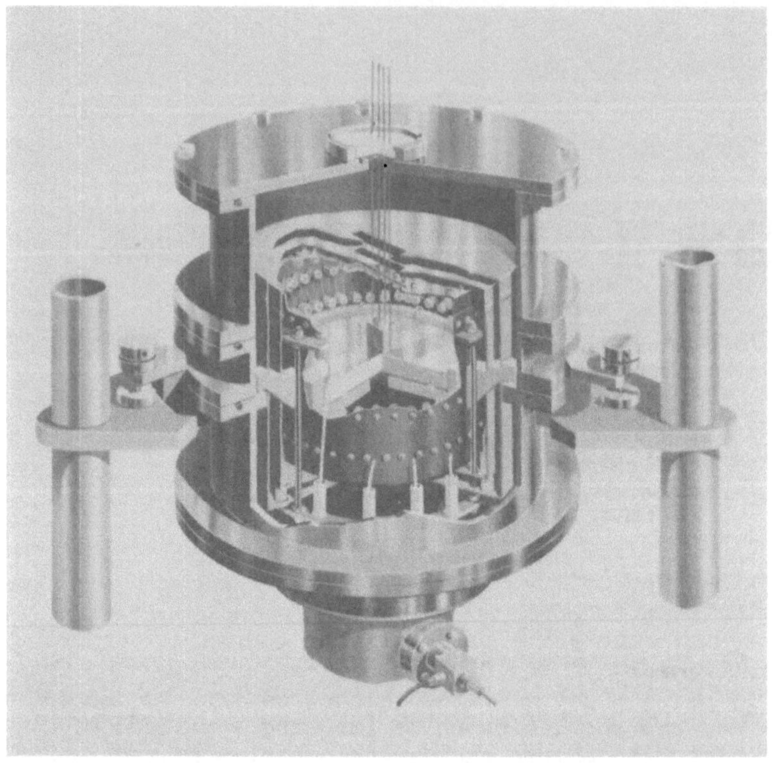

Figure 1. Sectioned view of the vacuum oven assembly.

tilt is achieved by double-passing the interferometer and inverting the wavefronts between passes so that any tilt introduced in the first pass is removed in the second.

The beam from a stabilized helium-neon laser is divided at beam-splitter 'a' into two beams which then traverse similar paths, one being reflected twice from the specimen, the other from the base plate. Beam-splitter 'b' is a polarizing multi-layer film which initially reflects the two beams into the oven. The quarter-wave plate below the interferometer rotates the plane of polarization of returning beams through 90° so that they are transmitted by the beam-splitter 'b'. The cube-corner returns the beams to the oven and also inverts the wavefronts. After the second pass, the beams are reflected by 'b' and recombine at beam-splitter 'c'.

Two detectors placed in the output beams generate electrical outputs which vary sinusoidally as the specimen expands. These signals are recorded continuously on a chart recorder for subsequent

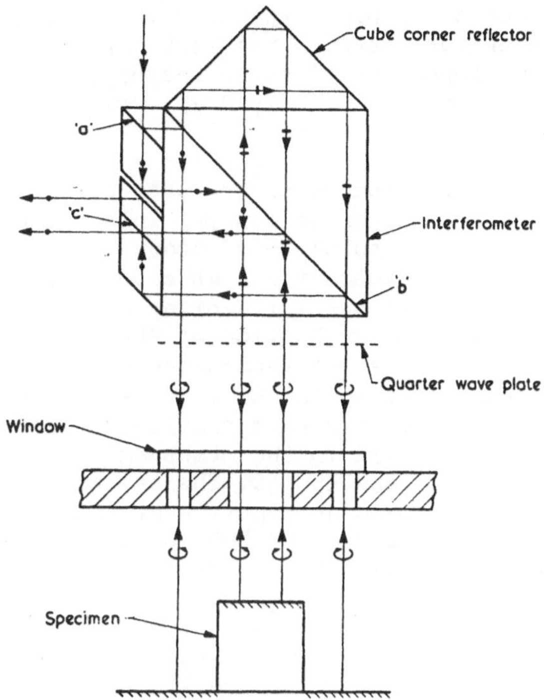

Figure 2. The double-passed interferometer.

examination and evaluation, and one cycle of interference corresponds to a quarter of a wavelength of expansion or contraction (approximately 158 nm).

ACCURACY

The total uncertainty in a measured expansivity is obtained as a combination of uncertainties associated with the measurement of change of length and change of temperature. In measurements on materials with low expansivities, the major contribution to the final uncertainty comes from the determination of the change of length. For instance, in the case of a 50-mm fused silica specimen measured over a 50-K interval, the total uncertainty due to laser wavelength variation and errors in determining the interference fringe fractions from the chart recording is estimated to be 0.8%, while the temperature measurement contributes 0.4%. For very low expansion materials, e.g. glass ceramics, the relative size of the ΔL uncertainties is even greater.

For metals and other high expansion materials, the measurement of change of temperature becomes the most important uncertainty source. If a 10-mm copper specimen is measured over 20-K intervals, the ΔL uncertainty amounts to only 0.2%, while the uncertainty in T contributes 0.7%.

Iron-constantan thermocouples are used with a data-logging system which has a resolution of 2.5 μV. This corresponds to a temperature measurement resolution of about 0.05 K, but these thermocouples, although chosen for high sensitivity, show some instability under repeated thermal cycling. They are all regularly calibrated by the Thermometry Section at NPL, and figure 3 shows the changes in calibration of three thermocouples of this type over periods ranging from eight months to five years. It can be seen that all three thermocouples demonstrate similar behaviour and would contribute errors up to 0.5% in the measurement of temperature intervals between 100 °C and 250 °C. The second recalibration of thermocouple Fe-Con 3 showed no significant change in this region, but some further instability above 400 °C. Some results for a platinum-platinum/13% rhodium thermocouple are also included in figure 3 for comparison. The better stability of these couples is, however, accompanied by a lower sensitivity. Used with a 1-microvolt voltmeter, they have a resolution of about 0.2 K at room temperature, improving to 0.1 K at 500 °C.

The overall stability (zero drift) of the instrument can be investigated by fully coating the base plate with a gold film and then operating the dilatometer without a specimen. All four reflections occur at the base plate, and no fringe movement should be detected when the oven temperature is increased. Three sources of error have been identified in this way and eliminated or reduced. Firstly, a vertical temperature gradient in the base plate produces a curvature of its upper face, which has been reduced by using a Cer-Vit plate and minimizing the vertical gradient. The second error is caused by thermal effects in the vacuum tank window, which have been reduced by water-cooling as much of the window surface as possible. Thirdly, the interferometer block can be affected by temperature changes, but the arrangement of figure 2 has been designed to make the path lengths in the glass equal so that no change of path difference occurs if the interferometer temperature changes uniformly. All these factors have been reduced to a level where they contribute a total path length change in the interferometer of only 0.3 wavelength when the oven temperature is changed from 20 °C to 500 °C.

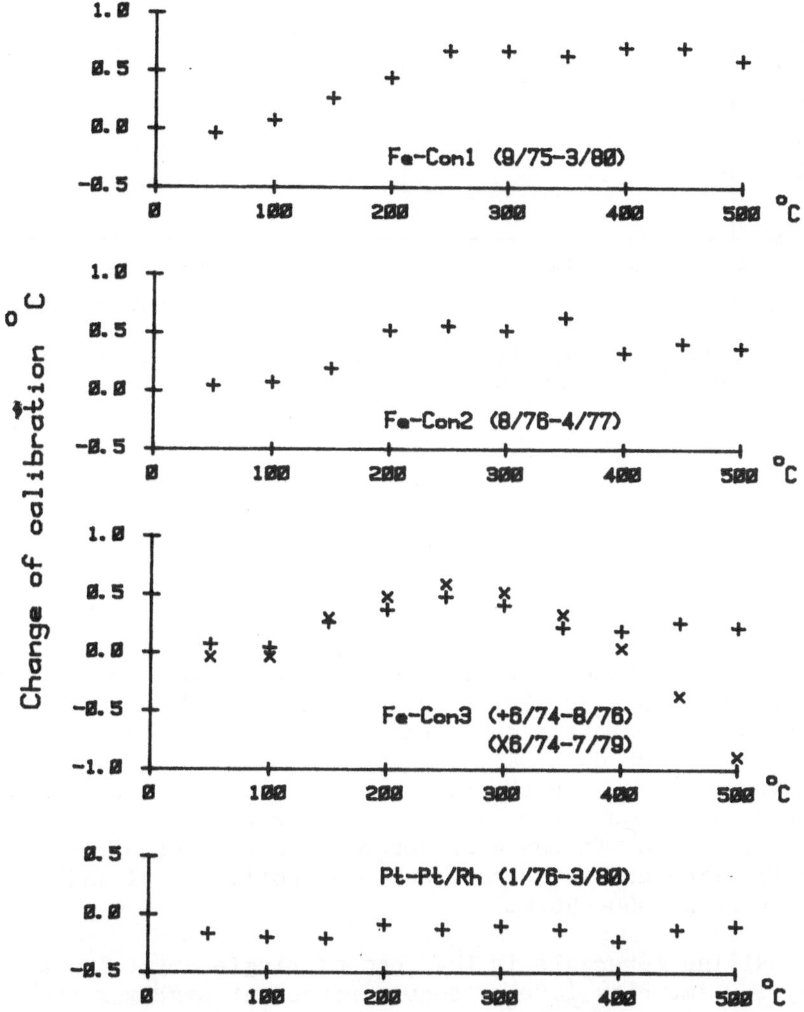

Figure 3. Changes of calibration of three iron-constantan thermocouples and one platinum-platinum/13% rhodium thermocouple after extended periods of thermal cycling.

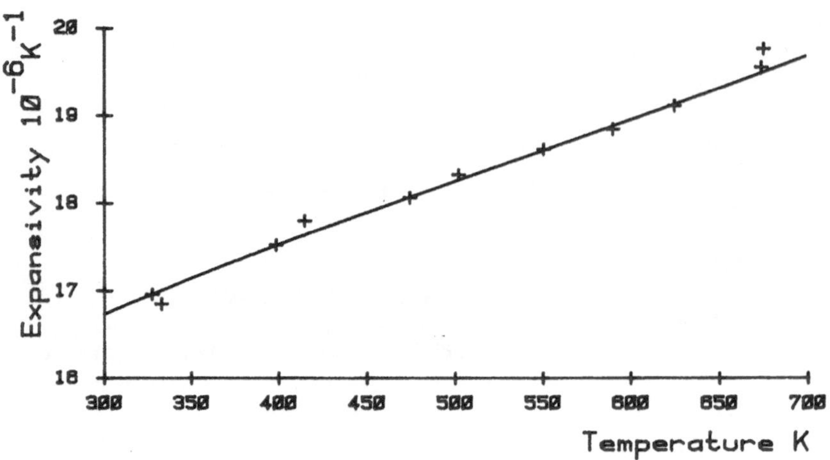

Figure 4. Measured expansivity of copper (SRM736) between 300 K and 700 K. The curve is the NBS calibration.

RESULTS

The NBS reference material, SRM736 copper, has been measured in the temperature range 300-700 K and the results are shown in figure 4. Each point represents a measurement of the mean expansivity over an interval of about 80 K, and the curve is the calibration given by NBS. All the points, with the exception of one at 675 K, lie within 1% of the calibration curve. A least squares third-order polynomial fit to the points of figure 4 produces a curve which agrees with the NBS certification and with the results of Kollie et al (1974) within 0.5% in the range 300-650 K.

Crystalline materials in the form of single crystals, available with very low impurity levels, should represent highly reproducible reference standards. Results are reported here for two materials, quartz (SiO_2) and silicon. The history of the crystals used for these measurements is not completely known, but the two quartz specimens were cut from a larger piece of quartz and have been optically examined to establish their orientation. The silicon is believed to be of high purity, having been part of a stock of material obtained in connection with a determination of Avogadro's number.

The quartz results, shown in figure 5, were obtained with two different specimens in a series of measurements of thermal expansion

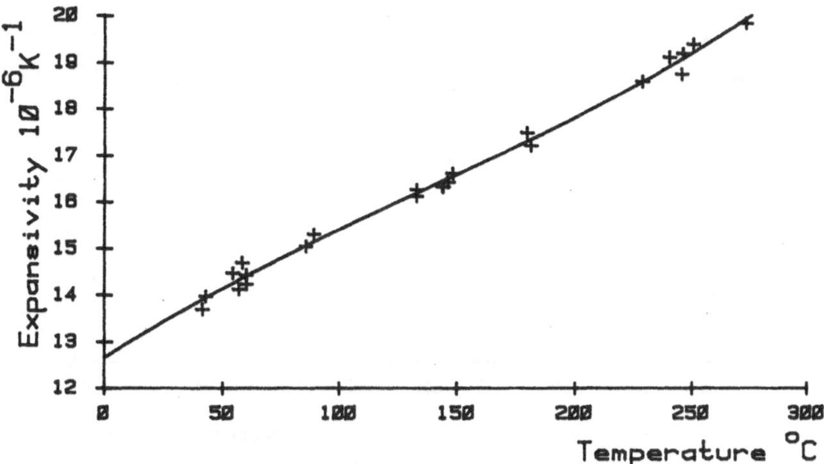

Figure 5. Measured expansivity of quartz normal to crystal axis, and third-order fit to the measured data (see text).

perpendicular to the crystal axis carried out between September 1973 and December 1978. Twenty three points have been plotted in the temperature range 20-300 °C, and the curve represents a least squares third-order fit to all the data, given by the polynomial

$$10^6 \times \alpha = 12.659 + 0.03266t - 6.844 \times 10^{-5}t^2 + 1.692 \times 10^{-7}t^3$$

The values calculated from this polynomial are about 1% lower throughout the range 20-300 °C than the results obtained by Amatuni and Shevchenko (1966). The origin of this difference cannot be positively identified and may be due simply to an accumulation of systematic errors in the two dilatometers used.

Figure 6 shows the measured expansivity of a silicon specimen. The results were obtained in two separate series of measurements over the temperature range 20-450 °C in approximate 40-degree increments. A third-order fit to the data is also plotted and is given by

$$10^6 \times \alpha = 2.410 + 0.00888t - 2.141 \times 10^{-5}t^2 + 2.266 \times 10^{-8}t^3$$

This polynomial gives values for the expansivity which agree within 1% with the results obtained by Novikova (1964) in the range 20-70 °C, and within 2.2% with the compilation by White (1979) throughout the range 20-450 °C.

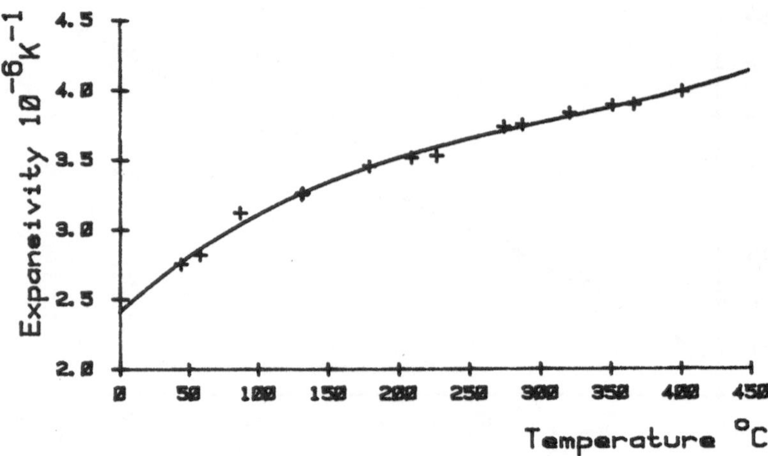

Figure 6. Measured expansivity of silicon and third-order fit to the measured data (see text).

CONCLUSIONS

The dilatometer described above is capable of making accurate absolute measurements of expansivity in the temperature range 20-500 °C. Its high stability and interferometric resolution make it particularly valuable for measurements on very low expansion materials where the greater part of the uncertainty of measurement is associated with the determination of change of length.

Measurements of the expansivities of a number of materials of general interest demonstrate the good reproducibility of the instrument. Agreement with other published results is generally within 1%.

Single-crystal materials (e.g. silicon), being readily available in a high state of purity, should be valuable as reference materials over a wide temperature range.

REFERENCES

Amatuni, A.N. and Shevchenko, E.B., 1966, Meas. Tech.,15:1256.

Bennett, S.J., 1976, J. Phys. E, 10:525.

Kollie, T.G., McElroy, D.L., Hutton, J.T. and Ewing, W.M., 1974, "Thermal Expansion 1973", AIP Conference Proceedings No. 17, pp 129-146.

Novikova, S.I., 1964, Sov. Phys. Solid State, 6:269.

White, G.K., 1979, High Temperatures - High Pressures, 11:471.

THERMAL EXPANSION OF IRON DURING THE $\alpha \to \gamma$ PHASE TRANSFORMATION BY A TRANSIENT INTERFEROMETRIC TECHNIQUE*

A.P. Miiller and A. Cezairliyan

Thermophysics Division
National Bureau of Standards
Washington, D.C. 20234

ABSTRACT

Measurements of thermal expansion of iron in the vicinity of (and during) the $\alpha \to \gamma$ phase transformation have been performed by a transient (subsecond) interferometric technique. The basic method involves rapidly heating the specimen from room temperature up to about 1300 K in less than one second by the passage of an electrical current pulse through it, and simultaneously measuring the specimen expansion by the shift in the fringe pattern produced by a Michelson-type interferometer and the specimen temperature by means of a high-speed photoelectric pyrometer. Results for the linear thermal expansion of iron at temperatures in ranges 1130 - 1180 K (α-phase) and 1200 - 1330 K (γ-phase) are expressed by the relations

$$\Delta\ell/\ell_o = -3.778\times10^{-3} + 1.480\times10^{-5}\ T$$

and

$$\Delta\ell/\ell_o = -1.883\times10^{-2} + 2.437\times10^{-5}\ T$$

respectively, where T is in K and ℓ_o is the specimen length at 20°C. The fractional change in length during the $\alpha \to \gamma$ phase transformation was determined, yielding 0.366%. The maximum inaccuracy in the measured expansion is estimated to be about 4%.

*This work was supported in part by the U.S. Air Force Office of Scientific Research.

INTRODUCTION

A few years ago, the development of an accurate high-speed interferometric technique was begun for the purpose of measuring thermal expansion of electrically conducting solids at high temperatures, primarily in the range 1500 K and the melting point of the specimen. This was accomplished by adapting a Michelson-type interferometer to the existing pulse heating system at the National Bureau of Standards [1,2]. Results for preliminary measurements on the thermal expansion of tantalum have been reported elsewhere [3], as have subsequent modifications and improvements to the interferometric system [4].

The basic method involves rapidly heating the specimen from room temperature to the maximum temperature of interest in less than one second by the passage of an electrical current pulse through it, and simultaneously measuring the specimen temperature by means of a high-speed photoelectric pyrometer [5] and the shift in the fringe pattern produced by the two-beam interferometer. The polarized beam from the He-Ne laser in the interferometer is split into two component beams, one which undergoes successive reflections from optical flats on opposite sides of the specimen, and one which serves as the reference beam. The linear thermal expansion of the specimen is determined from the fringe shift corresponding to a given temperature.

In our transient technique, the specimen is exposed to elevated temperatures for only a few hundred milliseconds thereby minimizing problems arising from heat loss, chemical reactions, evaporation, etc., which become especially severe in steady-state or quasi steady-state methods at temperatures above 2000 K. Another important characteristic of our techique is the capability of following thermal expansion of a solid continuously over an extended temperature range; this is of value when studying the behavior of substances exhibiting rapid changes in their thermal expansion, as in the case of solid-solid phase transformations.

The thermal expansion of iron is of particular interest because of a unique sequence of solid-solid phase transformations: as temperature is increased, the lattice changes from a body-centered cubic to a face-centered cubic structure ($\alpha \rightarrow \gamma$) at about 1190 K and then, at a somewhat higher temperature (~1660 K), returns to the body-centered cubic structure ($\gamma \rightarrow \delta$). In the present work, measurements of linear thermal expansion of iron in the vicinity of the $\alpha \rightarrow \gamma$ transformation were undertaken with the purpose of establishing the applicability of the transient interferometric technique to the study of thermal expansion during rapid solid-solid phase transformations.

The discussion related to instrumentation in this paper deals primarily with the operation of the two-beam interferometer. Detailed descriptions concerning the construction and operation of the pulse heating system [1,2] and the interferometric system [3,4] have been reported earlier.

INTERFEROMETER

The interferometer is a modified Michelson interferometer with the specimen acting as a double reflector in the path of one of the two beams, as illustrated schematically in figure 1. The specimen, shown in cross-section, is a long rectangular rod with two opposite sides polished optically flat.

The operation of the interferometer is relatively straight forward. A beamsplitter PB1 separates the linearly polarized beam from a He-Ne laser into two component beams. The component polarized normal to the plane of the interferometer serves as the

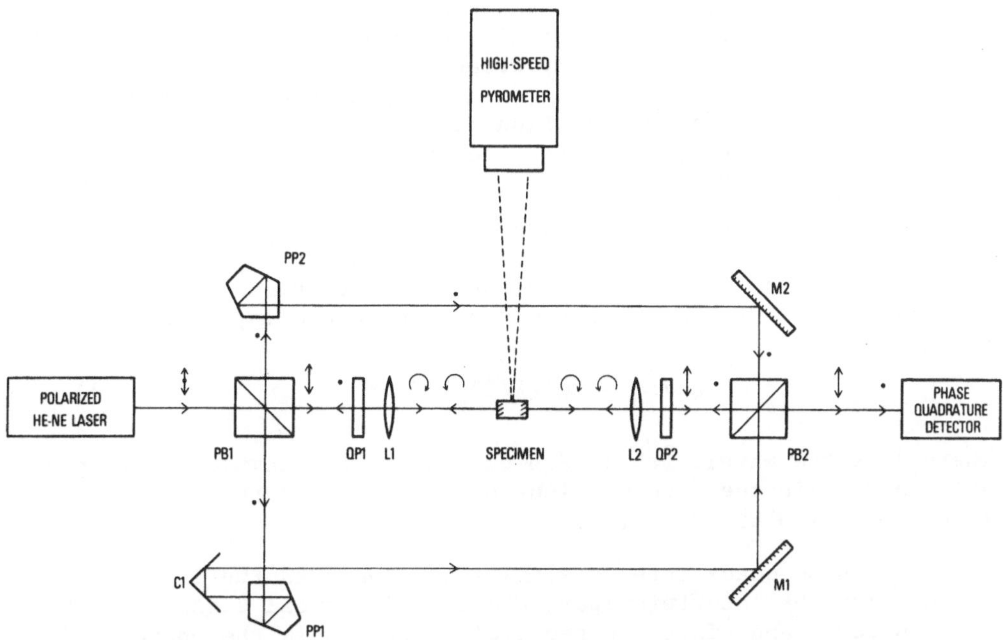

Figure 1. Schematic diagram of the interferometer consisting of optical elements: polarizing beamsplitters PB1 and PB2; quarter-wave plates QP1, QP2; lenses L1 and L2; corner cube C1; pentaprisms PP1 and PP2; and plane mirrors M1 and M2. The symbols ↕, •, ↷ and ↶ refer to polarization states of the component beams.

reference beam and therefore, is reflected around the specimen area and into the detector by the corner cube*/pentaprism combination C1/PP1, plane mirror M1 and a second polarizing beamsplitter PB2. The other component beam (its polarization parallel to the plane of the interferometer) is directed through quarter-wave plate QP1 (principal axis at 45° to the polarization planes) and then focused by lens L1 onto the polished "front" surface of the specimen where it undergoes reflection. A second traversal of QP1 rotates the polarization of the component through 90° causing the beam to be reflected around the specimen by PB1, pentaprism PP2 and mirror M2. By similar consideration of the optical elements PB2, QP2 and L2, one can show that after reflection from the "back" surface of the specimen the component beam enters the detector with its original polarization. The successive front surface/back surface reflections ensure that the optical path followed by this beam is insensitive to any translational movements by the specimen that may arise during rapid pulse heating.

Alignment of the interferometer is facilitated by placing an analyser (principal axis at 45° with respect to the polarization directions) in the path of the emergent beams; the position of the optical elements C1/PP1 and M1 are adjusted until the beams are superimposed, yielding very broad circular fringes. The contrast between the bright and dark fringes is maximized by rotating the plane of polarization of the input beam from the laser in order to compensate for unequal light losses along the paths of the two beams.

It can be shown that [3,4] a change in specimen "length" ℓ by one-half wavelength will give rise to a shift of one fringe through the field of view. Therefore, the fractional linear expansion of the specimen at a given temperature may be expressed as

$$\Delta\ell/\ell_o = \Delta n\ (\lambda/2\ell_o) \tag{1}$$

where λ is the wavelength (632.8 nm), Δn is the cumulative fringe shift and ℓ_o is the specimen length at a selected reference temperature (20°C).

The phase quadrature detector [3,4] converts the light output from the interferometer into two electrical signals which vary as cos δ and sin δ, respectively, where δ is the phase

---

*The polarization state of the beam transmitted by corner cube C1 is changed from linear to elliptical polarization; however, the original polarization of the beam is "restored" upon reflection by polarizing beamsplitter PB2.

difference between the two (orthogonally-polarized) light beams entering the detector. Fringe shifts are readily determined from changes in phase of either signal via the relation $\Delta n = \Delta\delta/2\pi$ where $\Delta\delta$ is measured in radians. For the purpose of the present study, sufficient information was obtained about the fringe movements by recording only one of the signals with a digital storage oscilloscope.

MEASUREMENTS

Analysis of typical material (nominally 99.95% pure iron) by the manufacturer yielded the following impurities (in ppm by mass): Aℓ, 60; Si, 50; S, 40; O, 33; Cr, Cu, Mo, 30 each; Pb, Sn, <30 each; C, 18; Ca, Mg, Ni, Ti, <10 each; Ag, <5; all other detected impurities, <1 each. Three specimens were fabricated from a polycrystalline rod into the form of long rectangular rods with nominal dimensions: 2.4 x 4.2 x 24 mm long; the specimen "length" for the interferometric measurements corresponded to the 4.2 mm dimension. Two opposite sides of each specimen were polished yielding a flatness of better than $\lambda/4$ across the center portions. The diameter of the light beam focused on the specimen surface was approximately 0.2 mm.

Prior to the pulse experiments, adjustments were made to the battery bank voltage and to a resistance in series with the specimen in order to achieve the desired heating rate (~2000 $K \cdot s^{-1}$). The specimen was then rapidly heated in a vacuum environment of about 1.3 mPa (~10^{-5} torr) from room temperature to the desired temperature in less than one second by passing an electrical current pulse through it. The maximum temperature achieved in a given experiment was determined by controlling the duration of the current pulse (~750 ms). The analog signals from the pyrometer and the interferometer were recorded approximately every 0.4 ms by means of digital storage oscilloscopes with a full scale resolution of about 1 part in 4000. In order to follow the very rapid fringe movements during the $\alpha \rightarrow \gamma$ transformation the interferometer signal was also digitally recorded every 20 μs with another digital storage oscilloscope. After each pulse experiment, the recorded data were transferred to a minicomputer for subsequent analyses.

Figure 2 presents three photographs, taken with successively expanded horizontal scale, of the interferometer signal as recorded during a typical pulse heating of iron through the $\alpha \rightarrow \gamma$ transformation; each cycle of the signal corresponds to the shift of one fringe or specimen expansion (contraction) of $\lambda/2$. The photographs show the reversal in direction of the fringe shifts at the onset of the $\alpha \rightarrow \gamma$ transformation, again at the completion of the transformation, and finally at the end of the pulse heating period.

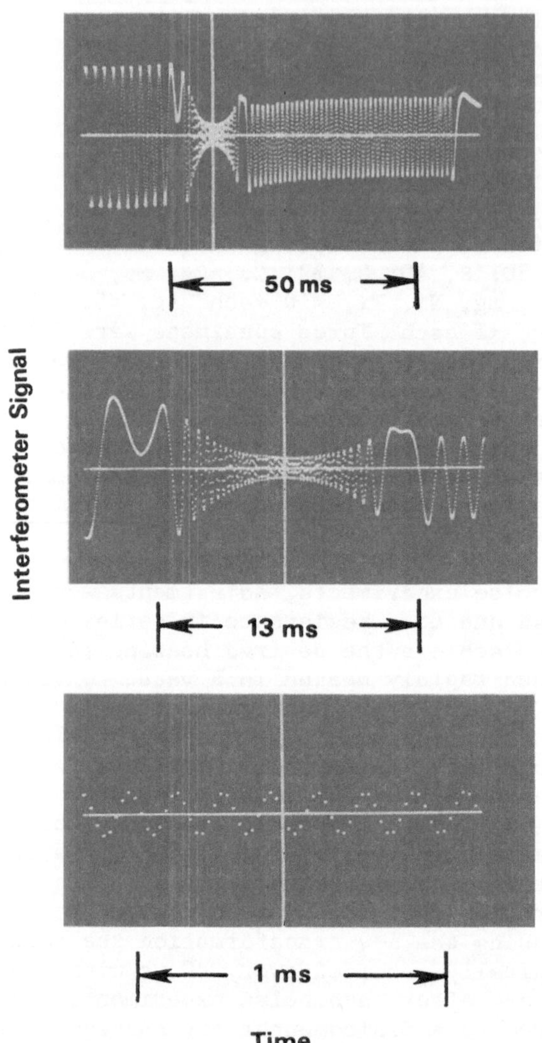

Figure 2. Three photographs of the interferometer signal as recorded during a typical pulse experiment in which iron is rapidly heated through the $\alpha \rightarrow \gamma$ transformation. The horizontal scale of the signal is expanded by factors of 1, 4 and 64 in the upper, middle and lower photographs, respectively.

The sinusoidal nature of the signal remains clearly visible throughout the transformation, even near the mid-point where the amplitude is significantly reduced. The rapid change in signal amplitude is probably related to the transformation kinetics associated with the transitory formation and movement of interphase "dislocation" boundaries at the specimen surface.

Upon completion of the experiments, we calibrated the high-speed pyrometer by means of a tungsten filament reference lamp which, in turn, had been calibrated against the NBS Photoelectric Pyrometer by the Radiometric Physics Division at NBS. All temperatures reported in this work are based on the International Practical Temperature Scale of 1968 [6].

RESULTS

The change in phase (δ) of the interferometric signal and, in turn, the fringe shift count Δn were computed for each specimen at time intervals of about 0.4 ms during the α-phase and 0.02 ms just before, during and after the $\alpha \to \gamma$ phase transformation. Figure 3 presents the fringe shift count as a function of time (only 1 out of 10 data points shown) during a typical experiment in which an iron specimen is pulse heated from room temperature up to temperatures beyond the $\alpha \to \gamma$ transformation. The wiggle in the fringe count versus time function in the vicinity of 600 ms reflects the combined effect of the "dip" in expansion coefficient of iron near its Curie point (~1040 K) and the rapid changes in the rate of temperature rise which result from the sharp peak in heat capacity of iron at the magnetic transformation.

The variation of specimen (true) temperature with time during the experiment is also shown in figure 3. Because of the lack in the literature of suitable normal spectral emittance data at 0.65 µm (the effective wavelength of the pyrometer), true temperatures were determined from the measured surface radiance temperatures* by selecting the "effective" normal spectral emittance ε_λ that would yield the correct transformation temperature at the arrest in the temperature-time function. The "correct" transformation temperature was selected to be 1190 K on the basis of data reported in the literature [7,8,9]. The "effective" value of ε_λ was assumed to remain constant for our computation of (true) temperatures throughout the range 1130-1330 K.

*Radiance temperature (sometimes referred to as brightness temperature) of the specimen surface is the temperature at which a blackbody has the same radiance as the surface, corresponding to the effective wavelength of the measuring pyrometer.

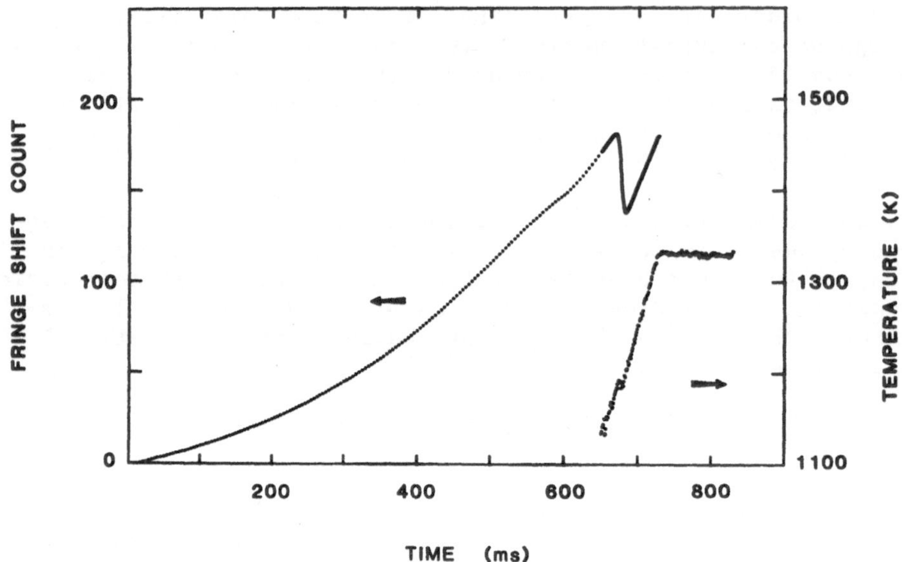

Figure 3. The fringe shift count and the specimen (true) temperature as functions of time during a typical experiment in which iron is pulse heated from room temperature up to temperatures beyond the $\alpha \rightarrow \gamma$ transformation.

Values of linear thermal expansion were obtained from the fringe shift count by means of equation (1) and correlated with measurements of the specimen temperature (determined approximately every 0.8 ms). Typical results for thermal expansion of iron in the vicinity the $\alpha \rightarrow \gamma$ transformation are illustrated in figure 4. The scatter in expansion values is primarily the result of random fluctuations in the temperature signal (see figure 3) which arise when operating the high-speed pyrometer below its optimum temperature range of 1500 K and above.

The final results for iron (based on a reference temperature of 20°C) were obtained by fitting the combined thermal expansion values for the three specimens by linear functions of temperature with the least-squares method. The function that represents the linear thermal expansion for α-iron (standard deviation = 0.8%) in the temperature range 1130 - 1180 K is

$$\Delta \ell / \ell_o = -3.778 \times 10^{-3} + 1.480 \times 10^{-5}\, T; \qquad (2)$$

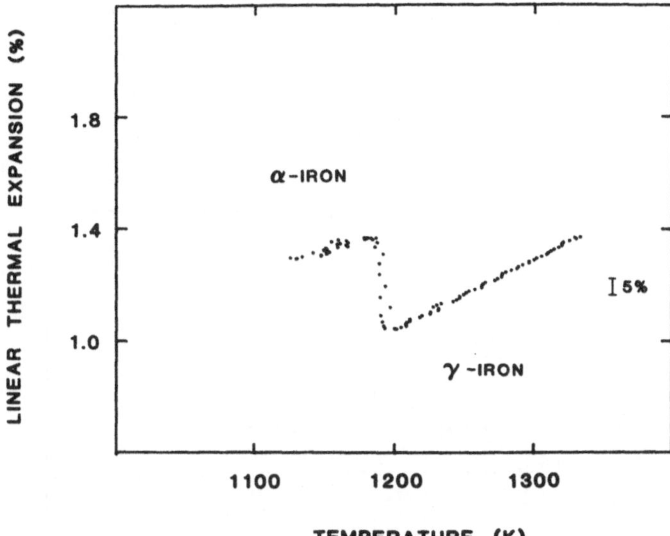

Figure 4. The linear thermal expansion of a given specimen of iron as a function of temperature in the vicinity of the $\alpha \rightarrow \gamma$ transformation.

for γ-iron in the temperature range 1200 - 1330 K, the function (standard deviation = 0.8%) is

$$\Delta \ell/\ell_o = -1.883 \times 10^{-2} + 2.437 \times 10^{-5}\ T \qquad (3)$$

where T is in K. The deviations of the results obtained for individual specimens from the overall fit represented by equations (2) and (3) are given in figure 5. The differences between the results for specimens 2 and 3 in the α-phase and specimens 1 and 2 in the γ-phase lie well within the random fluctuations in individual data points, whereas the results for specimen 3 in the γ-phase appear systematically lower than those for the other specimens by about 1%.

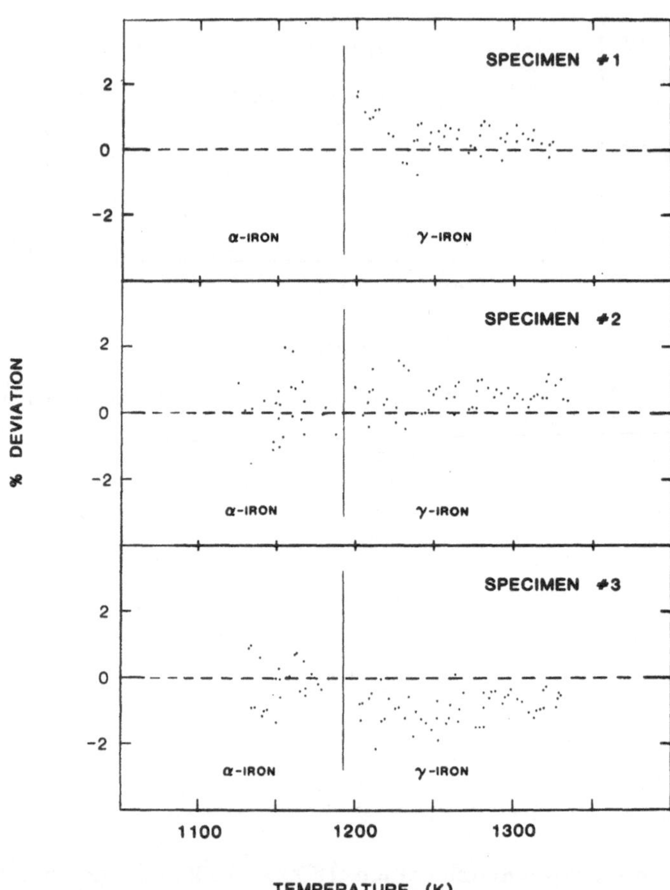

Figure 5. The percentage deviation of the values of thermal expansion for individual specimens from the overall fit represented by equations (2) and (3).

ESTIMATE OF ERRORS

The major source of error in our measurement of thermal expansion arises from the determination of specimen temperature. The uncertainty in selecting 1190 K as the transformation temperature in order to "anchor" the temperature scale in our measurements is believed to be about 10 K. The error in temperature, arising from the conversion of measured radiance temperatures into true temperatures under the assumption of a constant ε_λ, is estimated to be not more than 2 K on the basis of the very small temperature coefficients for ε_λ at values of λ near 0.65 μm [10]. A measure of the uncertainty resulting from the rather "noisy" pyrometer signal is given by the standard deviation (of an individual temperature measurement) obtained in fitting the measured temperatures to polynomial functions of time by the least-squares method; the standard deviation is approximately 4 K and 2 K for data in the α-phase and γ-phase, respectively. The effect of temperature gradients in the specimen on the measurement of temperature during rapid resistive self-heating has been considered in detail elsewhere [1]; under conditions of the present experiments, the error in temperature arising from temperature gradients is estimated to be negligibly small. Therefore, the maximum error in temperature, including all sources, is estimated to be about 16 K (α-phase) and 14 K (γ-phase) which correspond to errors in $\Delta\ell/\ell_o$ of approximately 2% (α-phase) and 3% (γ-phase).

The room temperature "length" (4.20 mm) of the specimen was determined to within 0.03 mm; this uncertainty contributes an error in $\Delta\ell/\ell_o$ of about 0.7%. The fringe shift count is believed to be accurate to within 0.5 fringe which corresponds to an additional error in $\Delta\ell/\ell_o$ of approximately 0.3%. It may be concluded that the maximum estimated error in the reported values of linear thermal expansion is about 4%.

DISCUSSION

The results of the present work are compared graphically in figure 6 with values of linear thermal expansion reported in the literature. Austin and Pierce [7], using Fizeau interferometry, observed considerable differences (~30%) in expansion among specimens of iron from several different sources; results for one of the "purest" specimens taken during the heating cycle of their experiment, are illustrated in figure 6 by open circles. The solid symbols represent expansion values as determined from the x-ray diffraction data reported by Basinski et al. [8], Gorton et al. [11] and Kohlhaas et al. [12] for iron samples of high purity (99.97, 99.67 and 99.97%, respectively). The values of

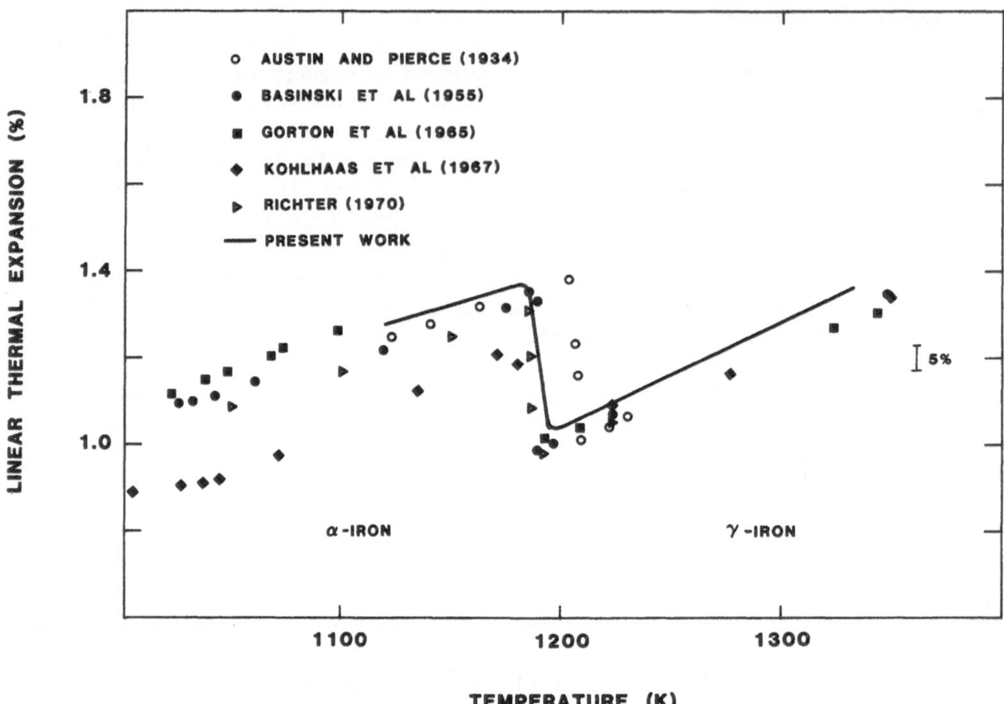

Figure 6. Linear thermal expansion of iron in the vicinity of the $\alpha \rightarrow \gamma$ transformation: present work and data reported in the literature.

thermal expansion reported by Richter [9] for 99.98% pure iron were obtained by conventional dilatometry; only data taken during the heating cycle are shown (open triangles).

From the results shown in figure 6, it is evident that the α-iron data of Kohlhaas et al. may be in error by perhaps as much as 20%. However, the agreement among the other data, including our results, lies well within the combined experimental error of the different investigations. A comparison of expansion values determined from length changes in bulk specimens (interferometry, conventional dilatometry) with values obtained from changes in lattice spacings (x-ray diffractometry) reveals no apparent bias with respect to measurement technique.

In the present work, the $\alpha \rightarrow \gamma$ transformation in iron was observed over a finite temperature range (~15 K) rather than at a constant temperature; this effect may partially be the result of impurities in the specimen, small changes in the normal spectral emittance of the surface during transformation, and/or small

Table 1. The fractional change in length for a bulk specimen of iron during the $\alpha \rightarrow \gamma$ phase transformation.

Investigator	Ref.	Year	Length Change (%)	Technique
Austin and Pierce	7	1934	0.25 - 0.37[a]	Fizeau Interferometry
Basinski et al.	8	1955	0.354[b]	X-ray Diffractometry
Richter	9	1970	0.33	Push-rod Dilatometry
Present work			0.366[c]	Pulse Interferometry

[a] Range of fractional length changes obtained among three of the "purest" specimens.

[b] Determined from changes in the lattice constant as iron is transformed from a body-centered cubic (α-phase) to a face-centered cubic (γ-phase) structure.

[c] Difference between values of $\Delta\ell/\ell_o$ obtained by extrapolating equations (2) and (3) to 1190 K.

temperature gradients in the specimen. Therefore, the fractional change in length during transformation was taken to be the difference between values of $\Delta\ell/\ell_o$ given by equations (2) and (3) at 1190 K, yielding 0.366%. From the results in table 1, it is evident that our value for the fractional length change in iron compares favorably with those obtained by investigators using other measurement techniques.

In conclusion, the results of the present study demonstrate the applicability of the high-speed interferometric technique to the measurement of thermal expansion during rapid solid-solid phase transformations. In future experiments, each specimen will be fabricated in the form of a precision machined tube (with optical flats) containing a small sighting hole for direct pyrometric measurement of the blackbody temperature. This will significantly reduce the uncertainty in measuring the specimen temperature and so, permit a more accurate determination of the linear thermal expansion of the specimen.

REFERENCES

1. A. Cezairliyan, J. Res. Nat. Bur. Stand. (U.S.) 75C, 7-18 (1971).

2. A. Cezairliyan, M.S. Morse, H.A. Berman and C.W. Beckett, J. Res. Nat. Bur. Stand. (U.S.) 74A, 65-92 (1970).

3. A.P. Miiller and A. Cezairliyan, in "Thermal Expansion 6", I.D. Peggs, ed. (Plenum Press, New York, 1978) p. 131-143.

4. A.P. Miiller and A. Cezairliyan, in "Thermal Expansion of Solids," to be published by CINDAS of Purdue University.

5. G.M. Foley, Rev. Sci. Instrum. 41, 827-834 (1970).

6. International Committee for Weights and Measures, "The International Practical Temperature Scale of 1968," Metrologia 5, 35-44 (1969).

7. J.B. Austin and R.H.H. Pierce, Jr., Trans. Amer. Soc. for Metals 22, 447-470 (1934).

8. Z.S. Basinski, W. Hume-Rothery and A.L. Sutton, Proc. Roy. Soc. (London) 229, 459-467 (1955).

9. V.F. Richter, Archiv Für Das Eisenhüttenwesen 41, 709-714 (1970).

10. Y.S. Touloukian and D.P. DeWitt, "Thermophysical Properties of Matter," Vol. 7, Thermal Radiative Properties, IFI/Plenum, New York, 1970.

11. A.T. Gorton, G. Bitsianes and T.L. Joseph, Trans. Met. Soc. AIME 233, 1519-1525 (1965).

12. R. Kohlhaas, Ph. Dunner and N. Schmitz-Pranghe, Zeit. Ang. Phys. 23 245-249 (1967).

A HIGH-SENSITIVITY POLARIZATION INTERFEROMETRY DILATOMETER

Y. Souche, R. Vergne, H. Ruby, J.C. Cotillard, J.L. Porteseil, G. Roblin,* and G. Nomarski*

Laboratoire Louis Néel, CNRS-USMG
166X 38042 Grenoble, France

*Institut d'Optique, Orsay, France

INTRODUCTION

A great number of optical dilatometers have been described using very often the advantages of interferometry. For example, let us recall Fizeau,[1] one of the first serious devices, made by Fabry and Arnulf, and some other well thought-of ones by research workers from the National Bureau of Standards (J.B. Saunders, G.E. Merritt, C.G. Peters, who published a long time ago in the Journal of Research of the N.B.S.). T.H.K. Barron, J.G. Collins, and G.K. White[2] and B. Yates[3] give an important bibliography about dilatometers.

Zeiss (Oberkochen) Company commercialized an end standards comparator. It is a double Michelson interferometer comparing two end standards, without any contact, by reflecting light on both faces. G. Roblin and Y. Souche[4] described a polarization interferometer applied to dilatometry in which they retained the interesting principle of the absence of mechanical contact with the sample. Elongations of the sample and of a reference length are compared, both at the same temperature. A.P. Miiller and A. Cezairliyan, more recently[5] proposed a Michelson type dilatometer applied to the transient interferometric dilatometer at high temperatures. Here, too, they eliminate every contact by reflecting light on both faces of the sample.

The present apparatus is, in fact, an improvement of the reference.[4] This dilatometer has been developed for the measurement of elongation during the phase transition of magnetic compounds. As the stresses can generate magnetic domains nucleations when the temperature passes through the phase transition we chose a method in which any mechanical contact is avoided. Moreover it is a polarization interferometer, so the changes of the path difference can be measured accurately. Indeed, in the better conditions, a linearly polarized vibration can be located within ±1", or $\pm 10^{-2}$ Å.

However, the present work is not pure metrology. In particular it has not yet been calibrated with reference materials but this operation would be done soon.

THE MEASUREMENT OF THE ELONGATION

Polarization interferometers of the Nomarski-Smith type have been used for a long time for the observation and measurement of transmitting or reflecting phase objects. Some years ago, we developed an application to dilatometry from which the present work is issued.[4]

The expansion of the sample is measured by comparison of its length L to the one of a reference sample L_0. So it is a differential device which converts the expansion $\delta(L - L_0)$ of the sample into an optical phase difference change $\delta\phi = 2\pi/\lambda \; 2\;\delta(L-L_0)$, then into the rotation of a linearly polarized vibration $\delta\theta = \delta\phi/2$. The measurement of the expansion of the sample is equivalent to tracking the azimuth of a linearly polarized vibration. An idea on the sensitivity of the measure is obtained by the correspondence between an expansion δL equal to $\lambda/2$ (halfwavelength) and $\delta\theta$ = 180°.

As any mechanical contact is avoided the optical beams need to reflect on the opposite faces of the sample and, consequently, of the reference length. Thus the interesting possibility of adding optical phases can be used (Fig.1). Indeed if a phase object Ω,

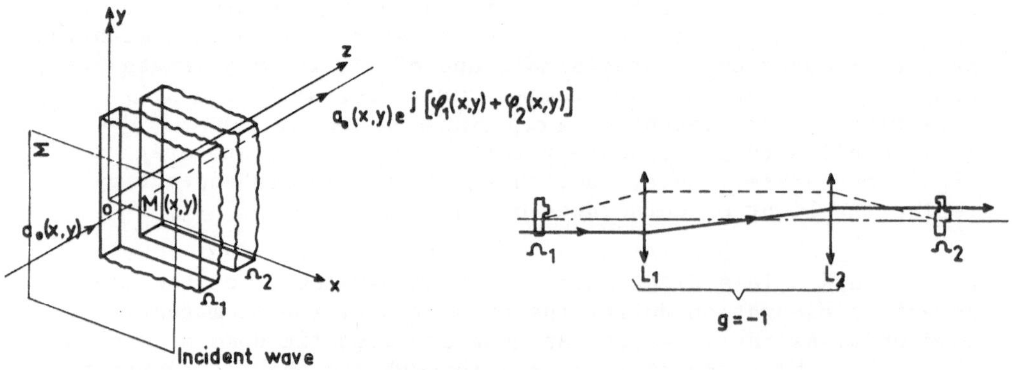

Figure 1. Principle of additivity of phases.

induces a path difference Δx_1 between two beams, another phase object Ω_2 on the same beam adds a path difference Δx_2 to Δx_1, so that the total path difference has become $\Delta x_1 + \Delta x_2$ as if the beams reflected only on one phase object equivalent to $\Omega_1 + \Omega_2$.

The highly polished sides of the sample and the reference form the phase object Ω_1, while the other faces correspond to Ω_2.

Fig.2 represents the paths of the optical beams in the instrument and Fig. 3 shows the arrangement of the set of the optical components. By means of a Wollaston W', the incident beam generated by an He-Ne laser splits up into two linearly polarized beams, the first of them being reflected by two faces of the sample, while the other is reflected by the reference sample. The Wollaston prism W makes the beams interfere.

The $\lambda/4$ plate of the Senarmont compensator provides a linearly polarized light crossed with the measuring analyzer.

The phase modulator between the polarizer P and the prism W' is a Jamin polarization interferometer adapted to modulation where one of the prisms is mounted at the top of a sinusoidally vibrating ferroelectric transducer.[5]

So the path difference before the $\lambda/4$ plate is resulting from the phase modulation (sinusoidal part) and from the thermal

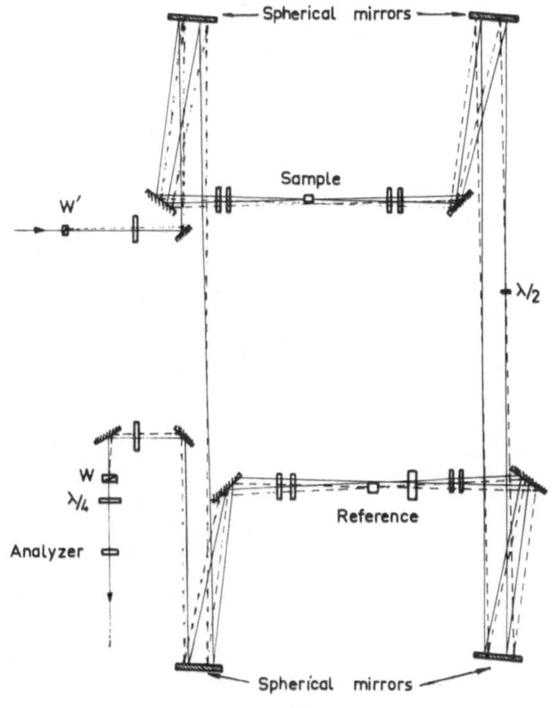

Figure 2

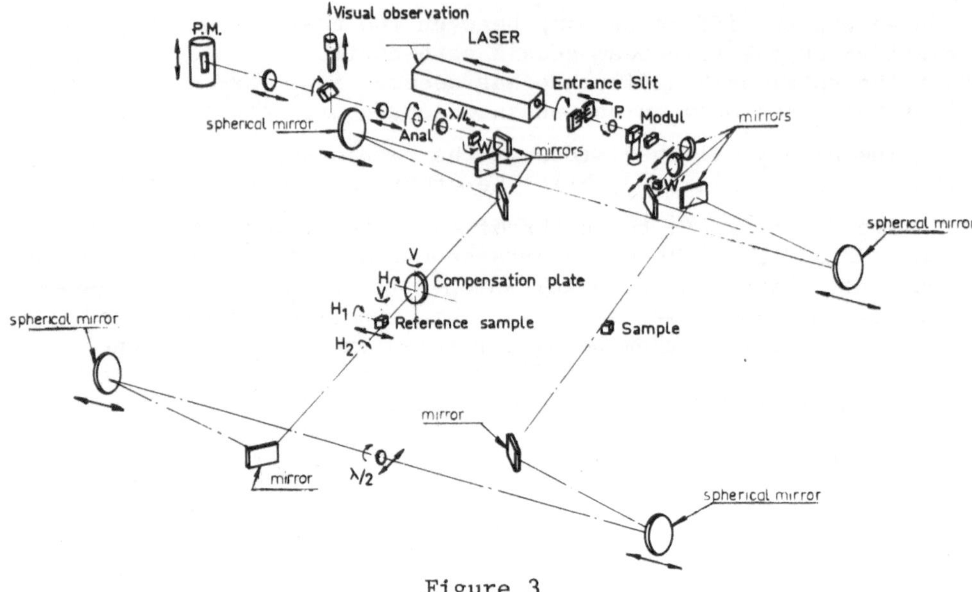

Figure 3

expansion (mean position of the phase). After the λ/4 plate the light is linearly polarized so that the mean position of the vibration is obtained by crossing the analyzer A. (Fig.4). A.D.C. motor rotates this polarizer while an optical encoder is measuring the rotations.

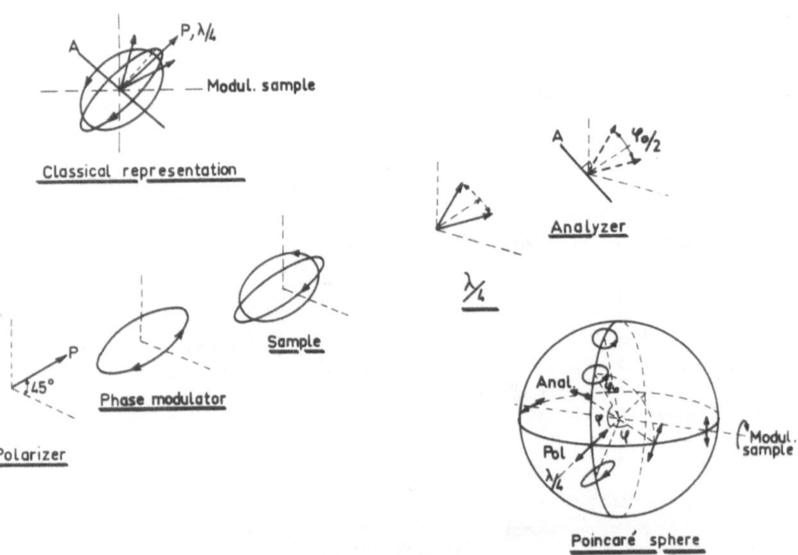

Figure 4

If ε is the error position of the analyzer, ϕ_0 the amplitude of the phase modulation and f_0 the frequency, the photoelectric signal S can be written as :

$$S \propto 1 - \cos(\varepsilon - \phi_0 \ \text{sing}\ 2\pi f_0 t)$$

A lock-in amplifier selects the amplitude of the term in f_0

$$S(f_0) \propto 2\ \varepsilon\ J_1(\phi_0)$$

So $S(f_0)$ is equal to zero when $\varepsilon = 0$. In fact all the odd components of the signal vanish when the error ε is cancelled.

The parallel glass plate (compensation plate) is used as a linear compensator for the optical paths introduced by the weak rotations of the sample during the measurements.

The half wave plate ($\lambda/2$) with its axes at $\pm 45°$ from the figure plane makes the optical device symmetrical with regard to the position of the polarization plans in relation to the incidence plans.

THE SAMPLE

The sample, the thermal expansion of which is measured, has some dimensional, optical, thermal and mechanical characteristics:

- dimensional: A 5 x 5 x 5 mm cubic sample has been scheduled. However, the length can be slightly different.
- optical: The two opposite faces of the sample and the reference are optically polished (1/2 fringe) and the parallelism must be the same as the reference sample (30" max).
- thermal: As some thermal expansion measurements require a very good temperature homogeneity in the sample it is necessary to ensure good thermal contact between the sample and its cooled holder in copper.
- mechanical: The sample must be held in a range from liquid helium to room temperature. However in the case of the study of the behaviour of solids near a transition point the sample must be free from stresses which would generate troubles for the critical behaviour. So the sample is held by a slight drop of glue (Silastene), the good properties at low temperature of which are well known.

PRODUCTION AND MEASUREMENT OF THE TEMPERATURE

The sample is located into an helium cryostat including a helium cooled exchange gas cell with optical windows. We have applied ourselves to work out two kinds of measurements:

- the usual measurements for which a very good resolution is not needed. The continuous temperature drift is used.
- the accurate measurements for which the temperature of the sample must be very well defined requiring a high quality for thermal regulation and a careful measurement for ΔT, the absolute measurement for T being usual.

For the first kind the measurement is automatized. The resistance change (C.L.T.S. or platinum or carbon) induced by the evolution of temperature is measured by means of the four-wire method connected with a digital multimeter.

The second kind corresponds to a step-by-step evolution so that a compensation method is used. The main difference with the first kind is that the measurement is achieved manually. Temperature is regulated with great care ($\pm 10^{-3}$K in better cases). The regulation is operated by a sophisticated proportional-integral-derival device with a large range of time constants.

PROCEDURE

When the sample temperature is drifting the measurement has to be done quickly. We describe briefly hereafter a typical sequence.

The analyzer A is positioned ahead of the azimuth of the mean position of the linearly polarized vibration issued from the $\lambda/4$ plate. This position is changing till it becomes perpendicular to the analyzer position. At that very moment, the computer picks up the interesting data (position of the analyzer provided by the angular encorder, temperature and gradient, sometimes the time interval), then it commands another step for the analyzer position still ahead of the phenomenon.

RESULTS

Elongation

The theoretical resolution corresponds to the one of the angular encorder, 2^{17} bis, that is 0.05 Å on one measurement.

For step-by-step measurements and samples in accordance with the above-mentioned requirements, resolution is 0.1 Å, that is, 0.2 Å for a stage with a time constant up to 1 sec. The elongation steps can amount to only a few angstroms.

Temperature

For temperature drifts resolution is about .02 K for each measurement while the accurate method gives .002 K for one step.

Compounds

Hereafter some results illustrate the possibilities of the apparatus.

The transition metal difluorides MnF_2,FeF_2 are tetragonal antiferromagnetic with the Néel temperature at 67,6 K and 78,3 K respectively. ZnF_2 is the isomorphous diamagnetic compound.

We present some results on rare-earth intermetallic ferromagnetic compounds $GdNi_5$, $TmNi_5$, $DyNi_5$, the thermal expansion of which is studied in collaboration with R. Lemaire and D.Gignoux research workers from Laboratoire Louis Néel, Grenoble. The changes of the behaviour between these compounds come from lattice terms and magnetic part including essentially crystal field effects.

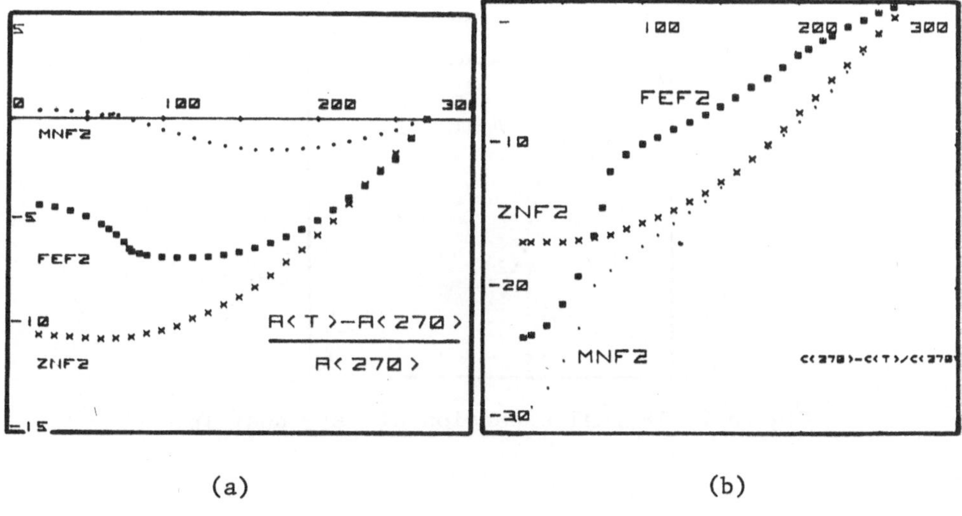

(a) (b)

Figure 5. Relative thermal expansion of difluorides.
(a) a axes, (b) c axes.

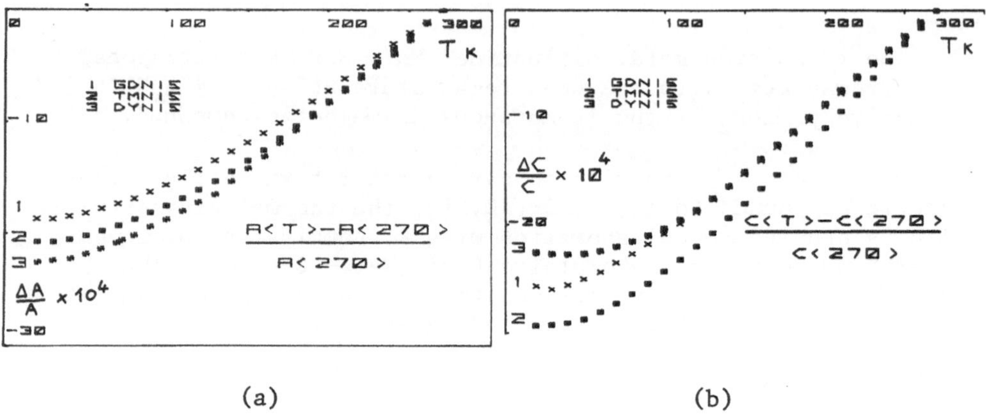

(a) (b)

Figure 6. Relative thermal expansion of $GdNi_5$, $TmNi_5$, and $DyNi_5$. (a) a axes, (b) c axes.

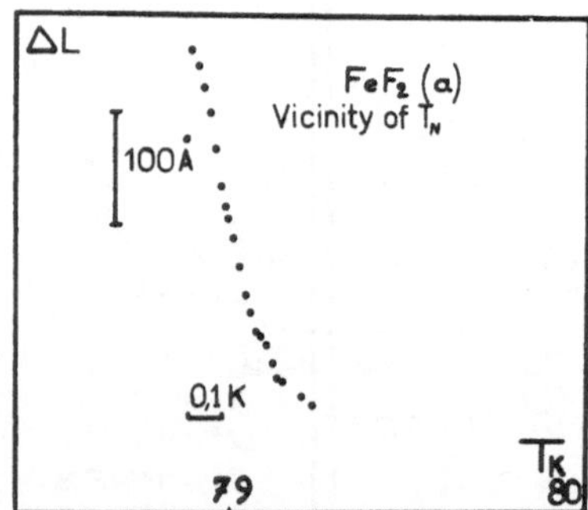

Figure 7. Thermal expansion of FeF_2 near T_N.

CONCLUSION

The apparatus that has been described is the first application of a true polarization interferometer to dilatometry. Moreover, the Jamin type modulator was mounted for the first time on an apparatus. G. Roblin[6] applied it, too, to the measurement of phase differences in microscopy.

The optical design has some interesting immunities on the one hand to the longitudinal motion of translation, to the rotation of the sample (geometrical insensitivity) and on the other hand to the gradients of air index as the sample and reference beams always travel very close to each other (about 2 mm).

Figure 8. General view of the interferometer.

REFERENCES

1. L. Fizeau, Ann.Phys. 128, 564 (1866)
2. T.H.K. Barron, J.G. Collins, G.K. White, Thermal expansion at low temperatures.Adv. in Phys., 29, 4 : 609 (1980).
3. B. Yates, Thermal expansion, Plenum Press - New York,(1972)
4. G. Roblin, Y. Souche, Interféromètre à polarisation destiné à la dilatoméjrie différentielle, Nouv.Rev.Optique, t.5, 5:287 (1974).
5. G. Nomarski, G. Roblin, Sur une méthode de modulation du retard optique, C.R.Acad.Sci. (Paris), 276B: 251 (1973).
6. G. Roblin, La mesure des déphasages en microscopie par interférométrie à modulation de phase. J. Optics (Paris), 8, 5 : 309 (1977).

EXPANSIVITY OF WATER UP TO 5 KBAR BETWEEN -28°C AND 120°C - EXTRAPOLATIONS IN THE SUPERCOOLED REGION I (DOMAIN OF ICE I)

Philippe Pruzan*, Léon Ter Minassian**, Alain Soulard***

*present address : Physique des Milieux Très Condensés
**Laboratoire de Chimie-Physique
***Laboratoire de Minéralogie-Cristallographie
Université Pierre et Marie Curie, 75230 Paris

SUMMARY

Through the measurement of the heat of compression, the expansivity α of water has been measured in the range 0-5 kbar for 14 isotherms lying between -28°C and 120°C. From a fit of $\alpha = \alpha(T,P)$ the isothermal compressibility κ_T and the specific heat Cp have been computed. Despite the fact that the measurements are generally made outside the supercooled region (Ice I domain) a fair description is obtained in this range.

INTRODUCTION

This work on the expansivity $\alpha = V^{-1}[\partial V/\partial T]_p$ of water is part of a continuing systematic study of the expansivity of materials.[1-5] The aim of the present study was twofold : first, to obtain an accurate description of the variation of α over a large p-T range and, second, to compute from the data other thermodynamic properties.

EXPERIMENTAL

The method of measurement, called the piezothermal method, has been fully described elsewhere.[1-5]

The heat exchanged between the sample and the thermostat during an isothermal compression or decompression is measured. The first Maxwell equation leads to α of the sample (Fig.1).[1-5]

RESULTS

Figure 2 shows the p-T range investigated. Below 0°C and at low pressures, measurements were performed only a short distance into the supercooled region (domain of stability of ice I). At high pressure the large supercooling effect afforded an easier access (domain of ice III,IV,V). Figure 3 shows that α is a decreasing function of pressure above 49°C (322.3 K), whereas below this temperature α increased with pressure (Fig.4). It is of interest to note that, as p decreases at low temperatures, there is a rapid decrease in α and a change in sign of $(\partial\alpha/\partial T)_p$.

FITTING OF THE DATA

We previously found that for a normal liquid (e.g., hexane)[5] the behavior of α vs p has a hyperbolic form, i.e.,

$$\alpha = a + b/(p + c),$$

where a, b, c are polynomials of degree 2 in temperature. A similar form is found here for water, but the function is more complicated and becomes

$$\alpha = a + b/(\pi + c)$$

where a, b, c are polynomials of degree, 2, 4 and 3, respectively, in temperature and π is a polynomial of degree 3 in pressure.[8] The mean value obtained for the relative residuals is of the order of magnitude of the uncertainty (1%). The general aspect of the isothermal is given in Fig.5. Figure 6 shows the shape of the isobars.

These data agree generally within 2% with other data computed from the volume measurement.[6-7]

COMPUTATIONS

The isothermal compressibility $\kappa_T = - V^{-1}(\partial V/\partial P)_T$ and the specific heat $C_p = T(\partial S/\partial T)_p$ have been computed from familiar formulas.

Comparisons with recent data[6-7] show a fair agreement (2% for κ_T, 5% for C_p)

Tables for κ_T and C_p are provided elsewhere.[8]

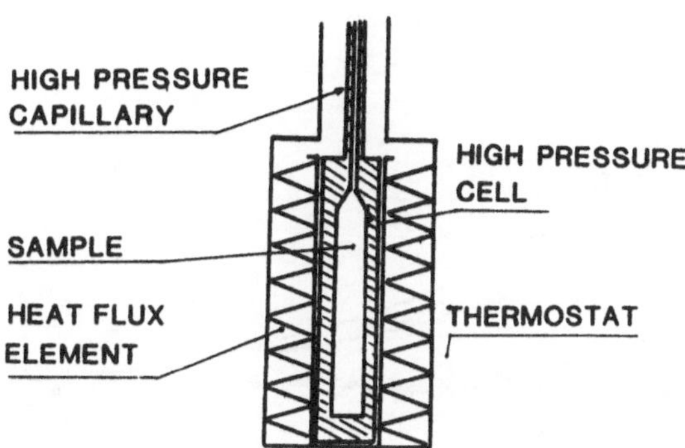

Fig.1. Principle of measurement of expansivity. The calorimetric element is sensitive to a definite volume V_e of sample. The calorimetric equation is written in the form $dQ = -(\alpha-\alpha r) VeTdp$. αr is the expansivity of the cell material.

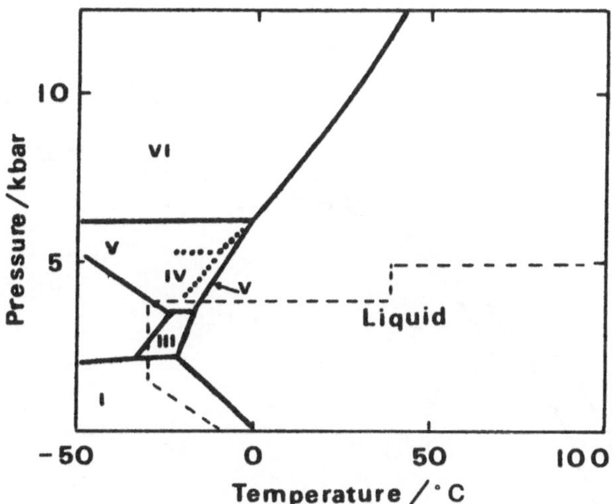

Fig.2. Phase diagram of water. The dashed line is the boundary of the range investigated.

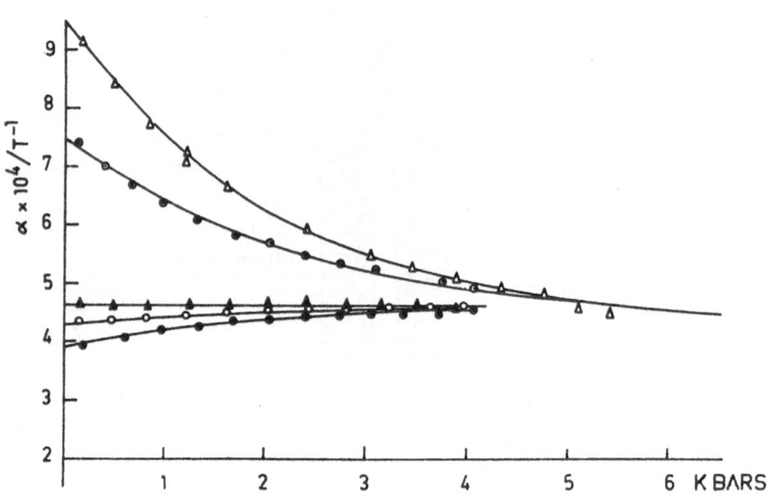

Fig.3. Expansivity of water. • 313 K, O 317 K, Δ 322 K, o 372 K, Δ 410 K.

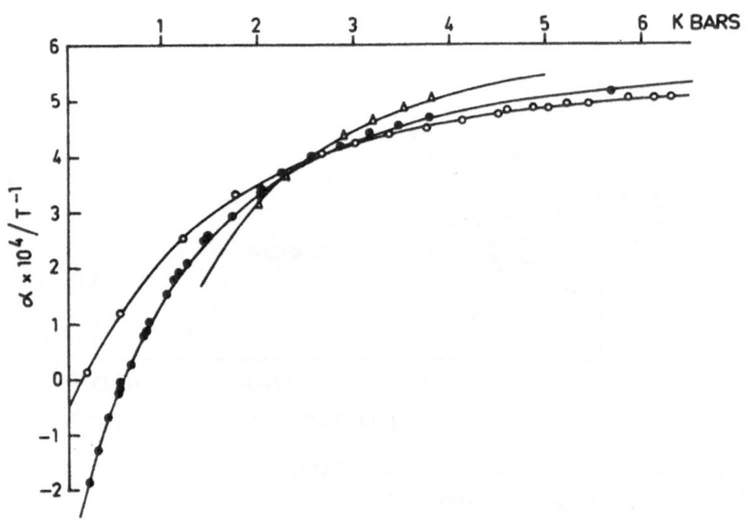

Fig.4. Expansivity α of water. Δ 246 K, • 263 K, o 275 K.

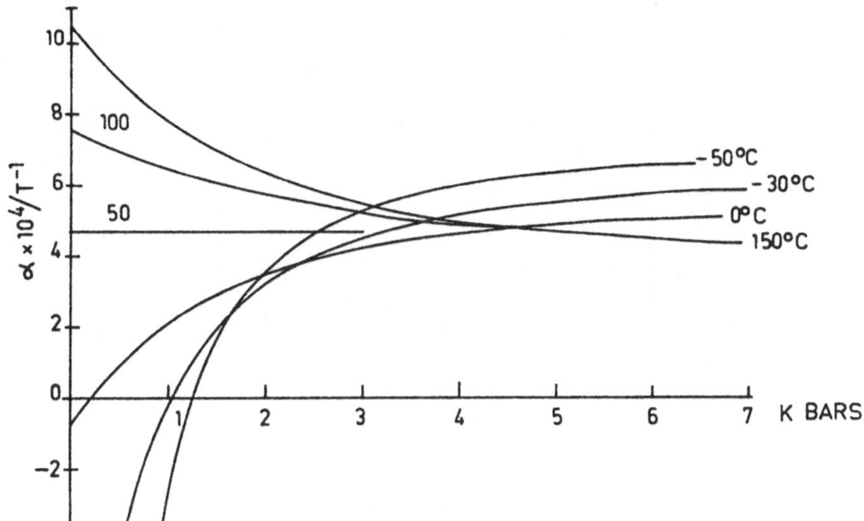

Fig. 5. Expansivity α of water. Isotherms are from fitted equation (1). Note the rapid decrease in α as p decreases at low temperatures and the reversal in sign of $(\partial\alpha/\partial T)_p$.

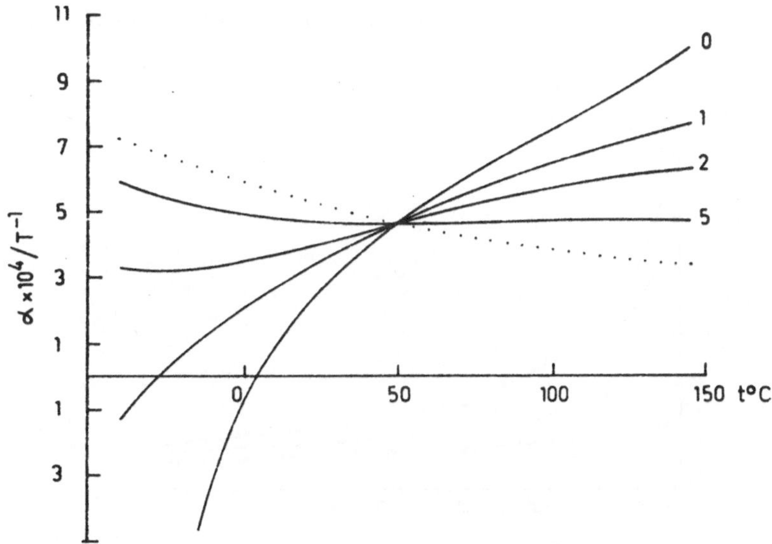

Fig.6. Expansivity α of water. Isobars are from fitted equation (1). Note the rapid decrease in α as T decreases at "low pressure" and the reversal in sign of $(\partial\alpha/\partial T)_p$. Dotted line : limit at infinite pressure.

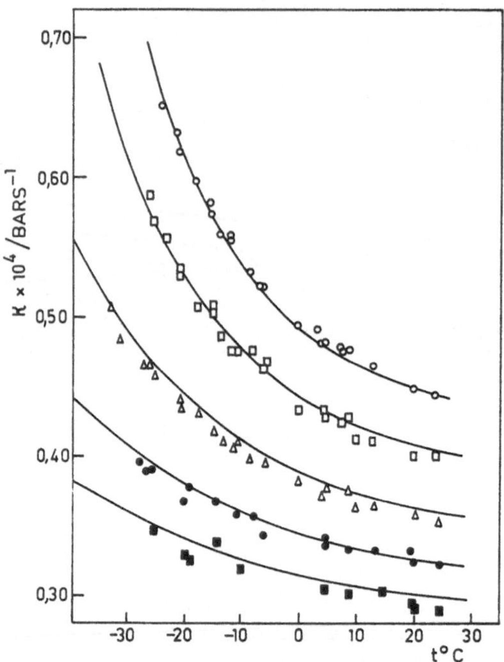

Fig. 7. Isothermal compressibility κ_T of water in the supercooled range. Data from (9); ○ 100 bars, □ 500 bars, Δ 1000 bars, ● 1500 bars, ■ 1900 bars. (critical exponent γ = 0.37) Solid line : computed from Eq.(1).(critical exponent γ=0.39)

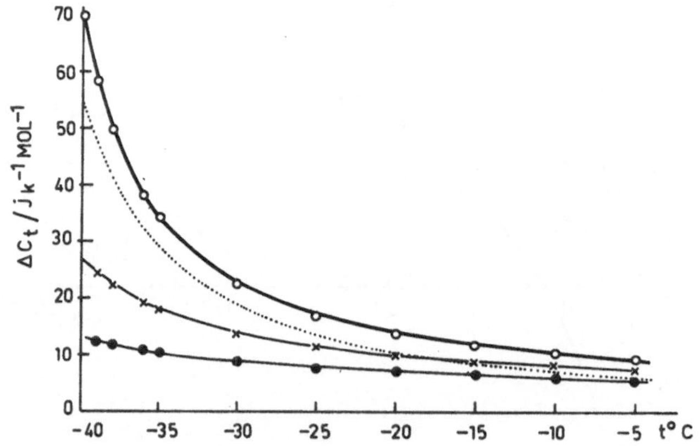

Fig. 8. Abnormal part of the heat capacity of water in the supercooled range estimated from this work, ○ 1 bar, × 200 bar, ● 400 bar (critical exponent ν = 0.91). Dotted line : fit from Oguni and Angell[10] (p = 1 bar, critical exponent ν = 1.35)

EXTRAPOLATION AT LOW TEMPERATURES

The very good accuracy of α , especially at low temperature justified extrapolation into the so called singularity range of water.

Behavior of $\alpha(T,P)$

Equation 1 (the set of isotherms or isobars Figs.5 and 6) shows that the asymptotes (parallel to the α axis) belong to the range $p > 0$ for sufficiently low temperatures. For example, for $t = -45.2°C$, the asymptote is found to be 1 bar. This fact may be associated with observation of the singularity in water. At atmospheric pressure other authors[9-10] place the singularity in the range -45°C, -40°C. The singularity moves toward higher pressures when T decreases.

Behavior of $\kappa_T(T)$

The isothermal compressibility increases sharply when T decreases. The extrapolation of our computation exhibits very good agreement with the data of Kanno and Angell [9] (Fig.7) - the critical exponent of $\kappa(T)$, that we found, is very close to that of these authors ($\gamma \simeq 0.38$).

Behavior of $C_p(T)$

The same features can be shown concerning the behavior of the abnormal part ΔC_t, of the specific heat, that we have estimated from our data. [8] Comparison is made with the fit of Oguni and Angell[10] obtained from their data (Fig.8, critical exponent $\nu \simeq 1.3$).

CONCLUSION

Accurate derivatives of state coordinates are required to get a fair description of thermodynamic properties. It is shown here that such data provide precise information around the boundary of the low temperature stability domain of water.

REFERENCES

1. L. Ter Minassian, Ph. Pruzan, J. Chem.Thermodynamics, 9, 375-390 (1977).
2. Ph.Pruzan, L. Ter Minassian, A. Soulard, in High Pressure Science and Technology, Vol.1, 368-378, (6th AIRAPT). Ed. K.D. Timmerhaus, M.S. Barber, Plenum (1979).

3. A.H. Fuchs, Ph. Pruzan, L. Ter Minassian, J. Phys.Chem.Solids, 40, 369-374 (1979).
4. L. Ter Minassian, Ph. Pruzan, J. Chem. Thermodynamics, 11 1123-1126 (1979).
5. Ph. Pruzan, L. Ter Minassian, A. Soulard, in High Pressure Science and Technology (7th AIRAPT) Vol.1, 226-229. Ed. B. Vodar, Ph. Marteau, Pergamon (1980).
6. T. Grindley, J.E. Lind, J. Chem.Phys. 54, 3983-3989 (1971).
7. Chen-Tung Chen, R.A. Fine, F.S. Millero, J. Chem.Phys., 65, 2142-2144 (1977).
8. L. Ter Minassian, Ph. Pruzan, A. Soulard, J. Chem.Phys. 75, 3064-3072 (1981).
9. H. Kanno, C.A. Angell, J. Chem.Phys., 70, 4008-4016 (1979).
10. M. Oguni, C.A. Angell, J. Chem.Phys., 73, 1948-1954 (1980).

LIST OF AUTHORS

Subject Index